Strategies for Creative Problem Solving

H. Scott Fogler

College of Engineering
University of Michigan
Ann Arbor, Michigan 48109-2136

Steven E. LeBlanc

College of Engineering
University of Toledo
Toledo, Ohio 43606-3390

For book and bookstore information

http://www.prenhall.com
gopher to gopher.prenhall.com

Prentice Hall PTR, Upper Saddle River, New Jersey 07458

Library of Congress Cataloging-in-Publication Data

Fogler, H. Scott.

  Strategies for Creative Problem Solving / H. Scott Fogler, Steven E. LeBlanc

    p. cm.

  Includes index.

  ISBN 0-13-179318-7

  1. Problem Solving.   2. Creative thinking.   I. LeBlanc, Steven E.   II. Title

BF449.F7 1994   94-26267

153.4'3—dc20   CIP

Editorial/Production Supervision: Lisa Iarkowski
Acquisitions Editor: Mike Hays
Manufacturing Manager: Alexis R. Heydt
Cover Design: Doug DeLuca

© 1995 by Prentice Hall, PTR
Prentice Hall, Inc.
Upper Saddle River, NJ 07458

Jolly Green Giant is a registered trademark of Green Giant Co., The Pillsbury Co.
Perrier Jouet is a trademark of Seagram's Chateau & Estate Wines Co.
Post-It Notes is a registered trademark of 3M Corporation
Pictionary is a registered trademark of Western Publishing Co., Inc. (Affiliate of Mattel, Inc.)
Shockblockers is a registered trademark of U.S. Shoe Corporation
Snickers, Milky Way, Mars Bar are registered trademarks of M&M/Mars, Inc.
Heath Bar is a registered trademark of Leaf, Inc.
The Franklin Day Planner is a registered trademark of Franklin Quest Co.
Coors is a registered trademark of Adolph Coors Co.

The publisher offers discounts on this book when ordered in bulk quantities.
For more information, contact:

  Corporate Sales Department
  PTR Prentice Hall
  113 Sylvan Avenue
  Englewood Cliffs, NJ 07632
  Phone: 201-592-2863
  FAX: 201-592-2249.

Printed in the United States of America

20 19 18 17 16 15 14 13 12

0-13-179318-7

Prentice-Hall International (UK) Limited, *London*
Prentice-Hall of Australia Pty. Limited, *Sydney*
Prentice-Hall of Canada, Inc., *Toronto*
Prentice-Hall Hispanoamericana, S. A., *Mexico*
Prentice-Hall of India Private Limited, *New Delhi*
Prentice-Hall of Japan, Inc., *Tokyo*
Pearson Education Asia Pte. Ltd., *Singapore*
Editora Prentice-Hall do Brasil, Ltda., *Rio de Janeiro*

TABLE OF CONTENTS

PREFACE

The purpose of this book is to help problem solvers improve their **street smarts**. We know that every individual possesses creative skills of one type or another, and that these skills can be sharpened if they are exercised regularly. This book provides a framework to hone and polish these creative problem-solving skills.

Strategies for Creative Problem Solving is for students, new engineers, practitioners, or anyone who wants to increase their problem-solving skills. After studying this book, the reader will be able to encounter an ill-defined problem, identify the real problem, effectively explore the constraints, plan a robust approach, carry it through to a viable solution, and then evaluate what has been accomplished. The skills to achieve these goals will be developed by examining the components of a problem-solving algorithm and studying a series of graduated exercises to familiarize, reinforce, challenge, and stretch the reader's creativity in the problem-solving process.

In order to cut through the maze of obstacles blocking the pathway to the solution to the problem, we need skills analogous to a pair of scissors with two special blades.

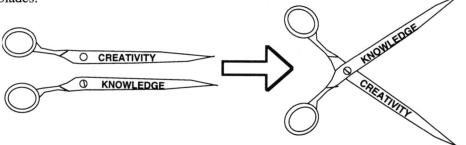

One of the shears is made of the knowledge necessary to understand the problem and to develop technically feasible solutions. However, no cutting can be done to solve problems of invention with just one shear. The other shear contains creativity that can generate new and innovative ideas. Creativity alone will not generate solutions that are necessarily technically feasible, and again no cutting can be done. Creativity along with a strong technical foundation, however, allows us to cut through the problem to obtain original solutions.

Over the past five years, we have researched problem-solving techniques used in industry. Teams of students and faculty have visited a number of companies (see acknowledgments) to study problem-solving strategies. We also carried out an extensive survey of new employees, experienced engineers, and managers in industry to collect information on the problem-solving process. As a result of our research, we know you can be a better problem solver.

A number of the engineers and managers provided examples of industrial problems that were incorrectly defined. These examples of ill-defined problems highlight the need to define the *real* problem as opposed to the *perceived* problem.

We believe that if a problem-solving heuristic had been applied to some of these problems in the first place, the true problem would have been uncovered more rapidly. *A problem-solving heuristic is a systematic approach to problem solving that helps guide us through the solution process and generate alternative solution pathways.* The heuristic in this book is quite robust and therefore applicable to many types of problems. However, we are not advocating the methods illustrated here as the *only* heuristic available; they are not. The problem-solving techniques presented in this book do, however, provide an organized, logical approach to generating more creative solutions.

The book is designed to lead the reader through the problem-solving process. Chapter 1 illustrates the need for an organized method of solving problems. Chapter 2 discusses the importance of approaching the problem with a positive attitude, the need for risk taking in the problem-solving process, and gives an overview of the heuristic (i.e., systematic approach). Subsequent chapters move step-by-step through the heuristic, shown below, to increase the reader's problem-solving *street smarts*.

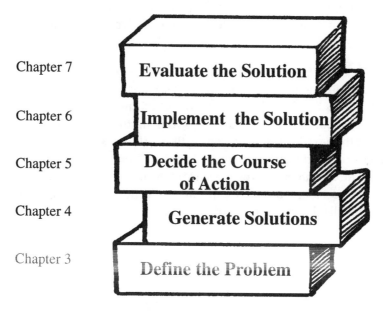

Chapter 7	**Evaluate the Solution**
Chapter 6	**Implement the Solution**
Chapter 5	**Decide the Course of Action**
Chapter 4	**Generate Solutions**
Chapter 3	**Define the Problem**

The Five Building Blocks of the Heuristic

ABOUT THE SOFTWARE

Interactive software that reinforces the concepts developed in the text is available for use on IBM compatible computers. Information on how to obtain the software for individual or class use may be obtained by writing *P.S. Interactive Software*, c/o CACHE Corporation, PO Box 7939, Austin TX 78713-7379, FAX: 512-471-7060, E-mail: cache@uts.cc.utexas.edu.

Eleven interactive modules are available to supplement and reinforce material in the text. The modules are:

- CONCENTRATION (Ch. 1 & 2) - An interactive puzzle/game that stresses the foundations of problem solving helps students learn the concepts.
- EXPLORE (Ch. 3) - This module reviews and exercises problem definition techniques as the student chooses the membrane for a heart-lung machine.
- DUNCKER (Ch. 3) - The student sharpens his/her problem definition skills by working on one of three scenarios involving vague problem statements.
- BRAINSTORMING (Ch. 4) - This module leads the student through a variety of brainstorming and blockbusting techniques.
- PLANNING (Ch. 6) - This module helps the student practice implementation of a problem solution using a student bridge building competition scenario.
- EVALUATION (Ch. 7) - As an employee in a paper mill, the student is asked to evaluate a proposed plant expansion.
- ETHICS (Ch. 7) - Ethical considerations are stressed in this module that finds the student as an engineer in a chemical company with an environmental problem.

Additional information is available on the World Wide Web site at http://www.engin.umich.edu/labs/mel/class/probsolv/pshome.html

ACKNOWLEDGMENTS

This book had its origins in *The Commission on Undergraduate Education* formed by Charles M. Vest, Dean of the College of Engineering at the University of Michigan. This commission's charge was to point the directions for engineering education for the next decade. It was during the workings of this commission that the need to focus on the development of students' problem-solving and creative skills was required more than ever before because of the increased global competition. To carry out the necessary research and data collection to develop strategies and materials to fulfill this need, a proposal was submitted to, and funded by, the National Science Foundation. This funding was instrumental in the conception and in the writing of this book.

We would like to acknowledge a number of people and organizations who helped make this work possible. We would like to thank Kepner-Tregoe, Inc. ("K-T") for its permission to use its copyrighted material in this book. Information on K-T short courses can be obtained by contacting K-T at P.O. Box 704, Princeton, NJ 08542. Telephone (609) 921-2806. We also wish to thank the following companies for participating in this project:

Amoco	Chevron Specialty Chemicals	Dow Chemical
Dow Corning	DuPont	Eli Lilly
General Mills	KMS Fusion	Kraft General Foods
Mobil	Monsanto	Procter & Gamble
Shell	3M	Upjohn

We would also like to thank the following faculty and students at the University of Michigan: S. Bike, R. Curl, C. Kravaris, J. Linderman, P. E. Savage, L. Thompson, and H. Wang; and A. C. Bushman, S. Bushman, J. Camp, P. Chen, D. Korotney, J. Gyenese, and J. Korniski, who helped collect, organize, and polish many of the problems in this book. Corinne Falender helped collect data on industrial problem-solving needs and examples. Michael Szachta participated in the preparation of a number of the figures as well as in the development of some of the industrial examples. A number of undergraduate students helped research and collect a number of the real-life examples used in the book by visiting shops, companies, and local merchants. These students are Chad White, Matthew Gdowski, David Graham, James Piana, Chris Teeley, Margaret Michael, Christina Nusbaum, Jen Casteel, and David Turczyn. Cathy Obeid, Susan Montgomery, Phillip Westmorleland, and Jeff Siirola gave a number of careful readings of the book and offered a number of helpful suggestions. Annette User proof read and corrected the copy edited version of the book, and also gave a final reading of the book, along with Michael Farnum, Matthew Gdowski, James Piana, and Christopher Domke. David Zinn provided most of the art work that appears throughout the text. In addition, Dr. Montgomery was a key participant in the development of the interactive software. H.S.F. would also like to thank his colleagues in the Chemical Engineering Department at Imperial College, London, and especially Julia Higgins and Stephen Richardson for their help and encouragement. We are also grateful to Professor Donald Woods for his pioneering work in bringing a structure of problem solving to the chemical and other engineering professions as well as initiating and stimulating the authors' interest in teaching problem solving. Janet Fogler spent endless hours editing and re-editing the book. Her comments and suggestions were invaluable to us. Last but not least we recognize Ms. Wendy Dansereau who helped prepare the initial versions of the manuscript, and Mrs. Laura Bracken, who typed and retyped what must have seemed like a never-ending succession of revisions as we converged on the final version. Their cheerful dispositions were always appreciated.

H.S.F.
Ann Arbor, MI

S.E.L.
Toledo, OH

July, 1994

1 PROBLEM-SOLVING STRATEGIES– WHY BOTHER?

Everyone is called upon to solve problems every day, from such mundane decisions as what to wear or where to go for lunch, to the much more difficult problems that are found in school or on the job. Most real-world problems have many possible solutions. The more complex the problem, the more alternative solutions there are. The goal is to pick the best solution. All of us will be better able to achieve this goal if we exercise our problem-solving skills frequently to make them sharper. By understanding and practicing the techniques discussed in this book, the reader will develop problem-solving *street smarts* and become a much more efficient problem solver.

1.1 WHAT'S THE REAL PROBLEM?

THE CASE OF THE HUNGRY GRIZZLY BEAR[1]

OR

AN EXERCISE IN DEFINING THE "REAL PROBLEM"

A student and his professor are backpacking in Alaska when a grizzly bear starts to chase them from a distance. They both start running, but it's clear that eventually the bear will catch up with them. The student takes off his backpack, gets his running shoes out, and starts putting them on. His professor says, "You can't outrun the bear, even in running shoes!" The student replies, "I don't need to outrun the bear; I only need to outrun you!"

The student realized that the bear would be satisfied when he caught one person; consequently the student defined the **real problem** as outrunning the professor rather than the bear. This example illustrates a very important point: *problem definition.*

Problem definition is a common but difficult task because true problems are often disguised in a variety of ways. It takes a skillful individual to analyze a situation and extract the real problem from a sea of information. Ill-defined or poorly posed problems can lead novice (and not so novice) engineers down the wrong path to a series of impossible or spurious solutions. Defining the "real problem" is critical to finding a workable solution.

Sometimes we can be "tricked" into treating the symptoms instead of solving the root problem. Treating symptoms (e.g., putting a bucket under a leaking roof) can give the satisfaction of a quick-fix, but finding and solving the **real** problems (i.e., the cause of the leak) are important in order to minimize lost time, money, and effort. Implementing *real* solutions to *real* problems requires discipline (and sometimes stubbornness) to avoid being pressured into accepting a less desirable quick-fix solution because of time constraints.

The next three pages present a number of real-life examples from case histories showing how easy it is to fall into the trap of defining and solving the wrong problem. In these examples and the discussion that follows, the *perceived problem* refers to a problem thought to be correctly defined but is not. These examples provide evidence of how millions of dollars and thousands of hours can be wasted by poor problem definition and solution.

Examples of Ill-Defined Problems

Impatient Guests

The Situation: Shortly after the upper floors of a high rise hotel had been renovated to increase the hotel's room capacity, the guests complained that the elevators were too slow. The building manager assembled his assistants. <u>His instructions to solve the perceived problem</u>: *"Find a way to speed up the elevators."* After calling the elevator company and an independent expert on elevators, it was determined that nothing could be done to speed up the elevators. Next, the manager's directions were *"Find a location and design a shaft to install another elevator."* An architectural firm was hired to carry out this request. However, neither the shaft nor the new elevator were installed because shortly after the firm was hired the real problem was uncovered. The **real problem** was to find a way to take the guests' minds off their wait rather than to install more elevators. The guests stopped complaining when mirrors were installed on each floor in front of the elevators.[2]

Leaking Flowmeter

The Situation: Flowmeters, such as the ones at the gasoline pumps to measure the number of gallons of gas delivered to your gas tank, are commonplace in industry. A flowmeter was installed in a chemical plant to measure the flow rate of a corrosive fluid. A few months after installation, the corrosive fluid had eaten through the flowmeter and began to leak onto the plant floor. <u>The instructions given to solve the perceived problem</u>: *"Find material from which to make a flowmeter that will not corrode and cause leakage of the dangerous fluid."* An extensive, time-consuming search was carried out to find such a material and a company that would construct a cost effective flowmeter. None was found. However, the **real problem** was to prevent the flowmeter from leaking. The solution was to institute a program of simply replacing the existing flowmeter on a regular basis before corrosion caused a failure.[3]

Bargain Prices

The Situation: A local merchant on Main Street in Ann Arbor was having difficulty selling a health food mix from the rain forest called Rain Forest Crunch, which was a hot selling item in other stores. Part of the attractiveness of Rain Forest Crunch was that it was indeed from the Brazilian rain forest and part of the proceeds of the sale went to protect the rain forest. The instructions given by the store manager: *"Lower the price of the item to increase sales."* Rain Forest Crunch still did not sell. The manager lowered the price further. Still no sales. After lowering the price two more times to a level that was well below the competitors', the item still did not sell. Finally, the manager walked around the store, and studied the display of Rain Forest Crunch. Then the real problem was uncovered. The problem was not the high cost of the item; the **real problem** was that it was not in a prominent position in the store to be easily seen by the customers. Once the item was made more visible, sales began to soar.[4]

| Price |
| Reduced |
| $14.99 |
| $12.99 |
| $10.99 |
| $9.99 |

Where Is the Oil?

The Situation: Water flooding is a commonly used technique in oil recovery in which water is injected into a well, displacing the oil and pushing it out another nearby well. In many cases, expensive chemicals are injected along with the water to facilitate pushing out the oil. A major oil company was having problems with a Canadian light-oil reservoir where the recovery was turning out to be much lower

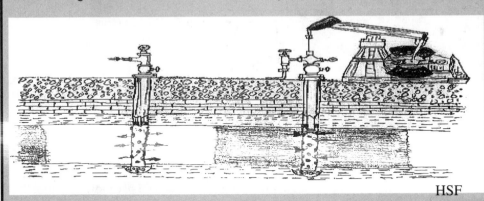

HSF

than expected. The instructions given to solve the perceived problem: *"Find ways to improve the oil recovery."*

Various studies costing hundreds of thousands of dollars were carried out over a 20-year period aimed at determining how to get more oil from the reservoir through improved water flooding techniques. Unfortunately, this situation wasn't a case of low oil recovery efficiency but rather one of miscalculation in the estimate of the amount of recoverable oil. In other words, there just wasn't much oil down there to recover![5] The **real problem** was to *learn **why** the well was not producing as expected* rather than ***how** to find ways to improve oil recovery.*

Making Gasoline from Coal

The Situation: A few years ago a major oil company was developing a process for the Department of Energy to produce liquid petroleum products from coal in order to reduce our dependence on foreign oil. In this process, solid coal particles were ground up, mixed with solvent and hydrogen, then passed through a furnace heater to a reactor that would convert the coal to gasoline (see figure below). After installation, the process was not operating properly. Excessive amounts of a tar-like carbonaceous material were being deposited on the pipes in the furnace, fouling, and in some cases plugging the pipes.

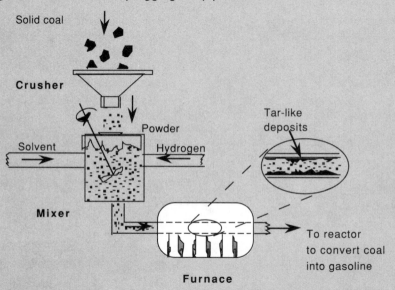

The instructions given by the manager to his research group to solve the perceived problem: *"Improve the quality of the solvents used to dissolve the coal and prevent these tar-like deposits."* A major research program was initiated. After a year and a half of effort was spent on the process, no one solvent proved to be a better solution to the problem than any other. Perhaps a more general problem statement such as, *"Determine why the carbon deposits are forming and how they can be eliminated"* would have revealed the true problem. The **real problem** was that the particles and solvent were reacting to form a coal-tar-like substance that was building up on the inside of the pipes in the furnace. The problem was solved by increasing the velocity through the furnace pipe, so that the particles and solvent had less time to react in the furnace to form the tar-like deposits. In addition, the high velocity caused the coal particles in the fluid to act as scouring agents on the furnace pipe wall. This velocity increase was accomplished by using a pipe of smaller diameter while maintaining the same total flow rate. After the furnace pipe was changed, no further problems of this nature were experienced.[6]

Better Printing Inks

The Situation. In 1990 the Bureau of Engraving and Printing (BEP) initiated a program to improve the quality of paper money being printed in the United States of America. <u>The instructions given to solve the perceived problem</u>: *"Develop a program to find better printing inks."* A number of workshops and panels were convened to work on this problem. After a year and a half of hard work by both government officials and college faculty on the perceived problem, research programs at several universities were chosen to try to develop better printing inks. Just as these programs were to be initiated, BEP withdrew the funds stating they had found that the **real problem** was not with the inks but with the printing machines. Consequently, the money earmarked for research on inks was diverted to the purchase of new printing machines. **By originally defining the wrong problem**, the Bureau of Engraving and Printing wasted thousands of hours of effort of government officials and college faculty.

Decreasing Profits

The Situation: In the 1980s, a government-operated factory in a developing country was taking material from a refinery and using it to make fertilizer. When the plant was designed and built, the price of the fertilizer was quite high and large profits were expected to be made. Unfortunately, shortly after the plant was in operation, the price of fertilizer dropped, and as a result the plant was operating at a loss. <u>The instructions given by the government to solve the perceived problem</u>: *"Close the plant because the price of the fertilizer is too low and we can no longer afford to operate it."* However, student engineers investigated the situation as a class project and found the **real problem** was not the price of fertilizer, but the inefficiency of operation which resulted from a power failure to the plant three to four times per week. There were enormous costs associated with restarting the plant after each power failure. The plant could still have made a substantial profit if the power failures had been prevented with emergency generators.

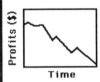

Time

1.2 RIGHT PROBLEM/WRONG SOLUTION

In this section, we discuss some examples where the real problem has been correctly defined, but the solutions to the problem were woefully inadequate, incorrect, or unnecessary. The persons who made the decisions in the situations described in these examples were all competent, hard-working professionals. However, some essential details that might have prevented the accidents and mistakes were overlooked. Using 20/20 hindsight, consider whether or not the following situations could have been avoided if an organized problem-solving approach had been applied.

Dam the Torpedoes or Torpedo the Dam

The Arcadian government wanted to increase agricultural production by finding ways to grow crops on waste lands. It was decided to cultivate land in the Pantoon region of southeastern Arcadia, which is very arid. Some wild plants could be seen growing in the soil from time to time, but there was insufficient moisture to grow crops. It was believed that the land could be irrigated and that agricultural food crops could be grown. The Orecha River, which flows naturally from the mountains to the sea, passes through the region. The solution chosen by the Arcadian government was: *"Design and build a dam to divert the river water inland to irrigate the land."* A multimillion dollar dam was built and the water diverted. Unfortunately, when the irrigation was achieved, absolutely no new vegetation grew, and even the vegetation that had previously grown on some of the land died. It was then determined that the infertility of the soil occurred because the diverted water dissolved abnormally high concentrations of salts present in the soil, which then entered the plant roots. Little of the vegetation could tolerate the salts at such high concentrations and as a result the vegetation died. A *Potential Problem Analysis* (Chapter 5) might have prevented this costly experiment.[7] Currently efforts are underway to deal with this salinity problem ranging from desalination to the construction of salt ponds.

DOWN

An Unexpected Twist

On June 1, 1974, the Nypro factory in Flixborough, England was destroyed and 28 men were killed when a vapor cloud of cyclohexane (a flammable chemical) ignited. Three units in the plant, each on a different level, were connected in a series. The middle unit was not operating efficiently. The problem statement was: *"Remove and repair the middle unit."* What followed was a faulty solution to the correct problem statement. When the middle unit was removed for repairs, a bent makeshift replacement pipe was used to connect the first and third units. A slight rise in the pressure and flow rates between the units caused the bent pipe to twist,

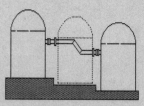 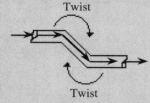

producing excessive strain. The pipe then ruptured, causing leakage of the vapor, which subsequently ignited, resulting in the explosion. Unfortunately, the only design consideration for the temporary pipe was a chalk drawing on the plant floor. Seemingly, the ramifications of such a replacement were not thoroughly thought out and certainly no concern was given to the strain that eventually caused the leak. This example shows the danger of a quick-fix solution that was not well thought out. Using the Kepner-Tregoe[8] technique of Potential Problem Analysis (KTPPA) might have prevented this disaster.

Carry out a KTPPA

The Kansas City Hyatt

The newly constructed Kansas City Hyatt Regency Hotel opened in 1980. It had three skywalks connecting the bedroom areas with the conference areas on all three levels. The skywalks were 120 feet long and were suspended from the roof.

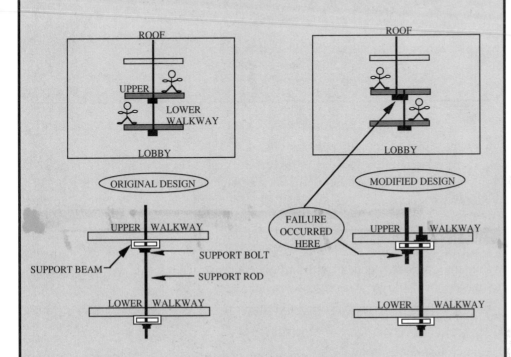

On July 17, 1981, a tea dance was being held in the lobby area, and people were watching from the skywalks and from the lobby below. The lower two skywalks collapsed, plummeting to the lobby below. Over 70 tons of concrete and girders fell to the ground. There were 114 people killed and hundreds more injured. It did not take long to determine the cause of the accident. In the original design, the top walkway was to be hung from the ceiling by long rods that passed through it and also supported the lower walkway. In this version of the design, each bolt had to support only one walkway. But somewhere between the original design and the actual construction, it was decided to replace each single long rod by two shorter rods. As a result, the bolt under the top walkway had to support not only the upper walkway, but the lower one as well, which doubled the force on the upper bolt. The connection failed when the bolt pulled through the upper walkway, and as a result, both walkways fell. A *Potential Problem Analysis* (PPA) on the modified design may have prevented the change from being implemented.

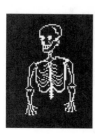

What's the Disease?

On a lighter note, we end with the following true example of Right Problem/Wrong Solution. At an American Medical Association (AMA) convention a number of years ago, an upper-body X-ray was displayed at the registration desk. The instructions given to the physicians as they registered : *"Diagnose the ailment from the X-ray, and place your answer in the contest box near the display"* (a correct problem statement). The winner of a valuable prize would be drawn from those who had made the correct diagnosis. Because of the focus on the upper torso, virtually every known lung disease was suggested by one physician or another. There was no need to hold a drawing from the correct diagnoses submitted because only one person discovered the true solution: Set a broken left arm.[9]

Nearly all project design failures, such as those above, result from faulty judgments rather than faulty calculations.

The goal of this book is to structure the process of defining and solving *real* problems in a way that will be useful in everyday life, both on and off the job. We shall achieve this goal by providing a structure to the problem-solving process called a *heuristic*. *A problem-solving heuristic is a systematic approach that helps guide us through the solution process and generate alternative solution pathways.* While a heuristic cannot prevent people from making errors, it provides a uniform, systematic approach to deal with any problem. In Chapter 2, we will continue discussing the use of a problem-solving heuristic.

> What's a
> Heuristic?

SUMMARY

Why bother with using a problem-solving strategy? This chapter presented a number of factual case histories that illustrate what happens when the real problem isn't defined or there is no organized approach to problem solving. In the chapters that follow, we will present a heuristic and a number of techniques that can greatly enhance the chances of defining and solving the **real problem** as opposed to the perceived problem and to identify **potential problems** during the design process.

REFERENCES

1. Prof. John Falconer, University of Colorado, Boulder, CO 80302.
2. Adapted from *Chemtech*, 22, 1, p. 24, 1992.
3. Dr. R.G. McNally, Dow Chemical Company, Midland, MI 48667.
4. Margaret Michael, University of Michigan, April 1993.
5. Dr. Mark Hoefner, Mobil Research and Development Corp., Dallas, TX 75387.
6. Prof. Antonio Garcia, Arizona State University, Phoenix, AZ 85287-6006.
7. True life example; only the country and other names have been changed.

8.   Kepner, C.H., and B.B. Tregoe, *The New Rational Manager,* Princeton Research Press, Princeton, NJ, 1981.

9.   Prof. Brymer Williams, University of Michigan, Ann Arbor, MI 48109.

FURTHER READING

Copulsky, William, "Stories from the Front," *Chemtech*, 22, p. 154, 1992. More anecdotal cases of histories of ill-defined situations and solutions.

EXERCISES

1.   Keep a journal of the ideas that speak to you as you go through this book. Begin by writing some thoughts below on what types of problems you would like to become more skilled in solving.

_____

_____

_____

_____

_____

_____

_____

_____

_____

_____

2.   Collect two or more ill-defined problems similar to the case histories described in this chapter.

_____

_____

_____

_____

_____

_____

_____

_____

2 GETTING STARTED

2.1 GETTING IN THE RIGHT FRAME OF MIND

Extensive research has been carried out on the differences between *effective* problem solvers and *ineffective* problem solvers.[1,2] The most important factors that distinguish between ineffective and effective problem solvers are the attitudes with which they approach the problem, their aggressiveness in the problem-solving process, their concern for accuracy, and the solution procedures they use. For example, effective problem solvers believe that problems can be solved through the use of heuristics and careful persistent analysis, while ineffective problem solvers think, "You either know it or you don't." Effective problem solvers become very active in the problem-solving process: They draw figures, make sketches, and ask questions of themselves and others. Ineffective problem solvers don't seem to understand the level of personal effort needed to solve the problem. Effective problem solvers take great care to understand all the facts and relationships accurately. Ineffective problem solvers make judgments without checking for accuracy. The table below further identifies differences between effective and ineffective problem solvers. *By approaching a situation using the characteristic attitudes and actions of an effective problem solver, you will be well on your way to finding the real problem and generating an outstanding solution.*

Characteristics of Effective/Ineffective Problem Solvers[1,2]

Characteristic	Effective	Ineffective
Attitude:	Believe the problem can be solved.	Give up easily.
Actions:	Reread the problem several times.	Lie back and hope a solution will occur.
	Redescribe the problem.	Unable to redescribe
	Ask themselves questions.	the problem.
	Create a mental picture.	
	Draw sketches, write equations.	
	Don't jump to conclusions.	Jump to conclusions.
Accuracy:	Check and recheck.	Do not check.
Solution Procedures:	Break the problem into subproblems.	Don't break the problem apart.
	Start at a point they first understand.	Don't know where to start.
	Use a few key fundamental concepts as building blocks.	Fail to identify key concepts.
	Use heuristics.	Guess.
	Persevere when stuck.	Quit.
	Use quantitative formulas, descriptions.	Do not do so.
	Keep track of progress.	Use no special format.

If you think
you can--
    you will.

If you think
you can't--
    you won't.

People who are effective problem solvers develop mind sets and habits which aid them in dealing with difficult problems. Stephen Covey's[†] research on highly effective people revealed that there are certain habits these people practice. The seven habits you should consider developing are shown in the table below.

The 7 Habits of Highly Effective People

Habit 1   **Be Proactive**. Take the initiative and make things happen. Aggressively seek new ideas and innovations. Don't let a negative environment affect your behavior and decisions. Work on things that you can do something about. If you make a mistake, acknowledge it and learn from it.

Habit 2   **Begin with the End in Mind**. Know where you are going and make sure all the steps you take are in the right direction. First determine the right things to accomplish and then how to best accomplish them. Write a personal mission statement describing where you want to go and what you want to be and how to accomplish these things.

Habit 3   **Put First Things First**. List your top priorities each day for the upcoming week and schedule time to work on them. Continually review and prioritize your goals. Say NO to doing unimportant tasks. Focus on the important tasks, the ones that will have impact if carefully thought out and planned.

Habit 4   **Think Win/Win**. Win/Win is the frame of mind that seeks mutual benefits for all people involved in solutions and agreements. Identify the key issues and results that would constitute a fully acceptable solution to all. Make all involved in the decision feel good about the decision and committed to a plan of action.

Habit 5   **Seek First to Understand, Then to Be Understood**. Learn as much as you can about the situation. "Listen, listen, listen." Try to see the problem from the other person's perspective. Be willing to be adaptable in seeking to be understood. Present things logically, not emotionally. Be credible, empathetic, and logical.

Habit 6   **Synergize**. Make the whole greater than the sum of its parts. Value the differences in the people you work with. Foster open and honest communication. Help everyone bring out the best in everyone else.

Habit 7   **Renewal**. Renew the four dimensions of your nature:
              Physical: Exercise, nutrition, stress management.
              Mental: Reading, thinking, visualizing, planning, writing.
              Spiritual: Value clarification and commitment, study and meditation.
              Social/Emotional: Service, empathy, self-esteem, synergy.

The upward spiral: Learn, Commit, Do; Learn, Commit, Do; Learn, . . .

It would be sad to work very hard to cut a path through a dense jungle only to find out it is the wrong jungle.

"Do not fear mistakes–fear only the absence of creative, constructive responses to those mistakes."

[†]*The 7 Habits of Highly Effective People*, COPYRIGHT©, 1989 by Covey, Stephen R. Reprinted by permission of Simon & Schuster, Inc., New York.

The previous table is meant to give only a thumbnail sketch of the 7 Habits, and the reader is referred to Covey's number one best-selling book for a more complete discussion, which includes a number of examples that illustrate these habits.

2.2 TAKING RISKS

Risks are actions, with little chance of succeeding, that require significant effort, resources, and/or time. However, if they are successful, they will have a major impact. Truly innovative solutions that make a significant difference in your life, organization, and/or community are almost never found without some risk taking. Although we don't advocate risk taking merely for its own sake, it can be an essential component to really creative solutions. Remember the old adage . . . the greater the risk, the bigger the reward.

Arsenio Hall was a successful department store manager. He gave up a "safe" job by taking a risk as a budding comedian. He failed a number of times before hitting it big, but we can safely say the reward was worth the risk!   The Dallas Cowboys was for years one of the most successful football franchises and Coach Tom Landry was one of the most respected field generals in the game. He was synonymous with the Dallas Cowboys. However in the late 1980s when the injury-laden Cowboys were having some unsuccessful seasons, the ownership changed. In a surprising move the new owner, Jerry Jones, fired the Dean of football coaches, Tom Landry and took a major risk by replacing him with Jimmy Johnson, who was an NFL rookie coach, and his college roommate. Within five years the Cowboys won two Superbowls, compared to only one Superbowl victory in the previous 25 years. There are many similar examples and we should take a lesson from them. Effective problem solvers have developed the proper attitude towards *risk taking*.

Ho Ho Ho

Another example of risk taking concerns the logo for Green Giant™ food products. "The Jolly Green Giant" first appeared as the symbol for Minnesota Valley Canning Company in 1925. However, when the company president proposed putting a green giant onto the label for canned peas, executives argued that it was *ridiculous* to have a giant with green skin. The executives were afraid to take a risk. "Whoever heard of green skin?" Fortunately, the president of the company was willing to take a risk. One could now ask the question "Who *hasn't* heard of the green giant?" He has appeared more than 450 million times on cans and been heard to say "Ho Ho Ho" more than 16,000 times in over 300 television commercials. (*NWA World Traveler*, 25, No. 3, p. 20, 1993)

There are some simple activities you can do that will make you become more comfortable with risk taking. For example,

- ask a question in a large lecture.
- go somewhere you have never gone before, (e.g., visit the Amazon).
- try a new sport, (e.g., skydiving).
- join a thespian group.
- volunteer to be the organizer of a group activity.
- take a challenging course outside your area of expertise.
- challenge established patterns of doing business in your organization.

Remember, to truly be considered a risk, an activity must have a chance of a negative outcome. To get in the habit of taking risks, you will need to develop a thick skin (i.e., not being too sensitive to criticism). Anytime you take a risk there will most likely be someone out there to criticize it.

Why Is Champagne Dry, Charles?

Up until the mid 1880s, champagnes were sweet and consumed at the end of a meal in much the same manner as ports and sherries. Charles Perrier was a successful champagne producer in France. In 1837 he began marketing and selling Perrier-Jouët™ Champagne in the United States, and between 1840 and 1870 exported over a million bottles. His success and fortune continued to grow to the extent that he constructed a $120,000 chateau (1870 dollars) in Epernay, France. The chateau featured six miles of underground cellars containing eight million bottles of champagne. In the mid 1880s, a family friend, John Crockfort, encouraged Perrier-Jouët to produce a dry (i.e., less sweet) champagne, one that would not compete with the after-dinner sherries and ports. Perrier-Jouët considered the idea and thought it was interesting. But why should they change? Who would buy it? Though not the leading champagne company in France, they were extremely successful and were concerned that such a change could bring ruination. Nevertheless, they did **take a risk** and began producing a dry champagne. Although it did take a while to catch on, by the early 1890s it was out-selling sweet champagne. By the turn of the century, over one million bottles per year of their dry champagne were being exported worldwide. Nowadays, virtually all champagnes are dry. (*NWA World Traveler*, 25, No. 8, p.28, 1993)

"We are making more progress by our failures than by our successes."
-*John Dunbar in "Dances with Wolves"*

The fear of failure is the greatest inhibitor to risk taking. When you are concerned about taking a risk, outline what the risk is, why it is important, and what would be the worst possible outcome. Next, describe what your options would be, given the worst possible outcome, and how you would deal with the failure. Learning how to deal with failure helps us to break away from the pattern of generating solutions that are "safe," but less than innovative. Failure accelerates the learning process by generating new information.

A course was recently offered in the University of Michigan's business school called *Failure 101*.[3] The basic premise of the course was to encourage risk taking by teaching the students not to be afraid to fail with the ideas they generated. The course provided many opportunities for students to fail on a number of projects in the marketing area. The class discussed examples of first failures that eventually developed into major successful ventures. The first pizza store of Tom Monaghan, owner of Domino's™ Pizza, went out of business. A glue the 3M company developed didn't stick well enough and was nearly abandoned until someone used it to develop Post-it™ notes. When the Petrossian brothers, who fled from Russia in 1917, introduced caviar at the Ritz Hotel in Paris, the French made ready use of the nearby spittoons. They were quite discouraged and could well have given up on the idea. Fortunately they persisted and overcame this first rejection, and today Petrossian Caviar is sold throughout France with prices up to $1,000/lb.

If major breakthroughs are to be made, risks will have to be taken. Failures resulting from these risks will occur but should not deter any future risk taking. The knowledge gained from these failures should be used constructively so that the chances of success will be even greater on the next try.[4]

2.3 LOOKING FOR PARADIGM SHIFTS

Joel Barker, in his work *Discovering the Future* speaks of the concepts of *paradigm shifts, paradigm paralysis,* and *paradigm pioneers*.[5] A paradigm is a model or pattern based on a set of rules that defines boundaries and specifies how to be successful at and within these boundaries. Success is measured by the problems you solve using these rules. Paradigm shifts can occur instantaneously or they can develop over a period of time. They move us from seeing the world one way to another. When a paradigm shifts, a new model based on a new set of rules replaces the old model. The new rules establish new boundaries and allow solutions to problems previously unsolvable. All practitioners of the old paradigm are returned to "ground zero" and are again on equal footing because the old rules no longer apply. For example, the guidelines (rules) followed by the most successful manufacturer of slide rules became useless by the paradigm shift in computation brought about by the invention of pocket calculators.

> *"... Boldly go where no one has gone before."*
> Captain Jean-Luc Picard

Barker describes *paradigm paralysis* as someone (or some organization) who is frozen with the idea that what was successful in the past will continue to be successful in the future. *Paradigm pioneers* are people who have the courage to escape a paradigm paralysis by breaking existing rules when success is not guaranteed. They realize that there are no easy roads when traveling in uncharted territory, and they cut new pathways, making it safe and easy for others to follow. The characteristics of a paradigm pioneer are the intuition to recognize a big idea, the courage to move forward in the face of great risk, and the perseverance to bring the idea to fruition. You need to be a paradigm pioneer, not only as you generate alternative solutions to a problem, but also as you look for ways to improve things when no apparent

problems exist. Additionally, paradigm pioneers should continually be searching for opportunities to initiate a *paradigm shift* to improve their process, product, organization, etc. Barker uses the example of the Swiss watch industry to make this point about paradigms.[5]

> ### A Paradigm Shift
>
> In 1968 the Swiss, with a respected history of making fine watches, held approximately 80% of the world market in watch sales. Today, they hold less than 10% of the market because of the emergence of the quartz digital watch. You will be surprised to discover, however, that the Swiss invented the quartz digital watch. A *paradigm shift* in wristwatch technology had occurred. The Swiss failed to adopt this new technology because they were caught in a *paradigm paralysis:* The idea that what was successful in the past will continue to be successful in the future. After all, "the digital watch didn't have a main spring, it didn't tick; who would buy such a watch?" Consequently, the inventors did not protect their invention with a patent, allowing Seiko of Japan and Texas Instruments (TI) to capitalize on the idea and market it. As a result of this paradigm paralysis, the employment in the Swiss watch industry dropped from about 65,000 to about 15,000 in a period of a little over three years. Even if the Swiss *had* decided to manufacture the digital watch after realizing its success, they would only have been on *equal footing* with Seiko and TI because of the paradigm shift. That is, all of their vast experience in making watches with gears and mainsprings would have given them absolutely no advantage in manufacturing digital watches.

Have you ever heard someone in your organization or business say "This is the way we have always done it. Everything seems to be going along OK, so why should we change what we are doing now?" If you have, you may have found someone caught in *paradigm paralysis*. If you recognize the symptoms of paradigm paralysis, you will be in a position to cure them by providing a **vision** and direction for your organization.

2.4 HAVING A VISION

Having a vision of the future is being able to see the way things ought to be (as opposed to the way they are now). It also includes a master plan for reaching this destination. It is imperative to identify a destination that is worthwhile reaching. A vision is essential for those who want to make a difference. George Bernard Shaw said: "Some people see things and say *why*, I see things that never were and say *why not*." Each one of us must look forward and find the voids in our organization, community, and life and try to fill them. We can achieve this by bringing together a coherent powerful vision through listening, reading, talking, and focusing our thoughts to find better ways of doing things. A vision with a master plan also makes day-to-day decisions easier by determining which of the decision choices is the most

> *"If you don't know where you are going, you'll probably wind up somewhere else."*
> --Yogi Berra

consistent with the master plan. We use our ethical and moral values to measure the rightness of our vision. However, vision is not the only thing we need.

> Vision without action is merely a dream.
> Action without vision merely passes the time.
> Vision with action can change the world.

To develop a vision, occasionally set aside a block of time (anywhere from a few minutes to several hours) to become introspective and to step back and look at the big picture. Determine what directions your life (or organization) should be taking, what needs to be accomplished, and devise a plan to meet your goals.

2.5 USING A HEURISTIC

In Chapter 1 we saw many examples of ill-defined and incorrectly solved problems. How can we avoid the same pitfalls as the people in these examples? The use of a problem-solving heuristic will help prevent many of these mistakes. A heuristic is a procedure that provides aid or direction in the solution of a problem. A heuristic is analogous to a road map. It can tell you where you are, where you want to go, and how to get there. Heuristics, like road maps, may also help you determine alternative routes to a destination. A complex problem, like a route selection, can be ill-defined, can have many choices, or possibly can have no feasible solution as posed. While there is no unique or preferred way to solve open-ended problems, we believe the use of a heuristic is an effective technique. The heuristic which we will be using is shown below and finds its origins in the McMaster Five-Point Strategy.[6] The complete McMaster Five-Point Strategy is given in Appendix 1.

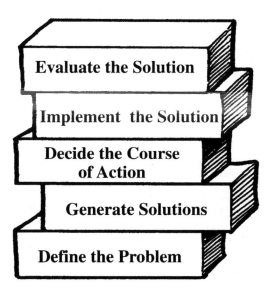

Figure 2-1. The Five Building Blocks of the Problem-Solving Heuristic

Building Blocks of the Heuristic

As you have seen in Chapter 1, the problem definition phase of the heuristic is very important. In practice, this phase can resurface at any point of the problem-solving process as one finds dead ends or changing criteria or conditions. Six techniques to help you arrive at the correct problem definition will be discussed in Chapter 3. We don't anticipate that you will use all six techniques all the time. In fact, different people will find some of the techniques more comfortable to use than others and the preferred techniques will vary from person to person. Once you have *defined the real problem*, it is then important to ask the following questions: Has this problem been solved before? Is it worth solving? What resources (time, money, personnel) are available to obtain a solution? Have you collected all the necessary information by reviewing data, reading the literature, and talking to colleagues and those involved in the problem?

After gathering significant information about the problem, you can proceed to the next step in the heuristic, which is to *generate* alternative solutions (Chapter 4). One of the most popular techniques used in industry to generate ideas is brainstorming. This technique is useful in expanding thinking as to what *is* possible and what *is not* possible. In addition to brainstorming, other methods to facilitate idea generation include analogies and blockbusting.

After you generate a number of solution alternatives, the next step in the solution process is to *decide which alternative to choose* (Chapter 5). Here, logic and analysis of each alternative are major factors in reaching a decision. Once the decision is made, you need to plan to ensure its success by identifying things that could go wrong, the causes of each potential problem, the preventive actions that could be taken, and the steps of last resort.

Having made the decision and planned for its success, you may now *implement* the solution (Chapter 6). The first step is to plan the activities you need to do to solve the problem. A number of techniques to allocate time and resources are presented to carry the solution through to successful completion.

In the *evaluation phase* (Chapter 7), you need to look back and make sure all of the criteria in the problem statement were fulfilled and that none of the constraints were violated. Has the problem *really* been solved, and is the solution the best solution? Is the solution innovative, new and novel, or is it merely an application of principles (which, in some cases, may be all that's necessary)? Is the solution ethical, safe, and environmentally responsible? Although *evaluation* is listed at the end of the heuristic, you should also evaluate the problem solution at various points along the way, especially when major decisions are made or branch points occur.

2.6 FOSTERING CREATIVITY

Apple Computer is perceived as being one of the more creative companies of the past decade. John Sculley (former chairman of Apple Computer) discussed the philosophy of maintaining a creative environment for product development.[7] Some of the ideas that he suggests for team leaders or managers to foster a creative environment are shown in Table 2-1.

TABLE 2-1: Establishing a Creative Environment[7]

"Don't give people goals; give them directions (i.e., roughly aim them).*"*

"Encourage contrarian thinking." Dissent stimulates discussion, prompting others to make more perceptive observations. It ultimately influences decision making for the better.

"Build a textured environment to extend not just people's aspirations but their sensibilities." You can't buy creativity, you can inspire it. Creative people require an atmosphere conducive to thinking in nonstandard ways. The work environment needs to be informal and relaxed.

"Build emotion into the system." Defensiveness is the bane of all passion-filled creative work. One way to keep defenses down is to encourage problem-finding as well as problem-solving. The world is moving so fast that problems are being created all the time. The people who can find them have tremendous powers of creative observation.

"The safer you make the situation, the higher you can raise the challenge." The workplace should be safe, so that the workers are not afraid to take risks and make mistakes, but the standards should be set high.

"Encourage accountability over responsibility." Traditional responsibilities, like punching a clock from nine to five, can inhibit creativity. Instead, people are made accountable for the results of their work.

"Getting ordinary people to reach beyond themselves and do extraordinary things can be the result of establishing a nurturing creative environment." To establish this type of atmosphere requires the reconciliation of traditional corporate attitudes with more iconoclastic entrepreneurial attitudes. If this fine line can be traversed without abandoning the very characteristics that have led to a company's success in the past, much progress can be made.[8]

It is essential to set the proper atmosphere for creativity to flourish. One of the ways to do this is being an effective leader who inspires others and leads by example. There are distinct differences between leaders and bosses, as summarized in the following table.

Characteristics of Leaders and Bosses

A Boss...		A Leader...
demands respect	but	earns respect
is a taskmaster		is a coach
is critical		is encouraging
rules by fear		guides by example
commands		inspires
makes work a burden		makes work fun
punishes mistakes		rewards success

Dora Dodge[8] effectively sums up these difference in her short poem "The Boss," which ends with "The boss gets compliance. The leader gets committment." Everybody needs leaders, but nobody needs a boss.

2.7 INTERACTING CREATIVELY

The ideas presented by Sculley for establishing a creative working environment are complemented by Scholtes who discusses the necessity for creative interactions with all people.[9] He makes the point that these interactions are essential for quality leadership. The principles of quality leadership stem from the premise that everyone you interact with at your workplace should be treated as a **customer** whom you must strive to satisfy: The customer may be a neighbor, supervisor, subordinate, co-worker, or an external client. Some of the leadership principles Scholtes specifically identifies are

- Customer focus
- Obsession with quality
- Continued education and training
- Unity of purpose

Scholtes urges us to *listen*, *listen*, and *listen* again to the *customer*. Find out what the customer's needs are. Involve the customer in the problem-solving process. Brainstorm possible solutions with the customer. Get input from the customer at each step and make him or her an integral part of the process. Relentlessly pursue the best quality product and/or service. Provide a structure where everyone can continue to learn so they can not only upgrade their skills but also keep at the forefront of their fields. The continual upgrading of one's skills is especially important in today's competitive marketplace, because many companies no longer guarantee employment based solely on loyalty and years of service. Rather, companies look at an employee's current skills and how these skills will be useful to the company. Have *everyone* on your team committed to work together to strive for excellence in everything they do.

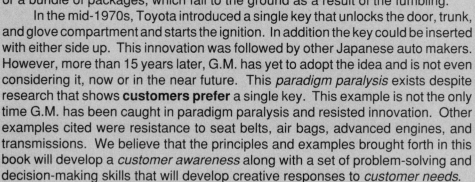

The *Key* Is Listening to Your Customers

A recent article in the <u>New York Times</u> (reprinted in the <u>Detroit Free Press</u>, p. B1, January 3, 1993) states that a major reason consumers have defected from purchasing automobiles from G.M. is because they compare the key entry system and the key ignition of the G.M. car with that of a Toyota car. The G.M. cars require two keys: one that can be used only for the car door and the trunk while the other one can be used only for the ignition. The idea behind this development is that you could allow someone to drive your car without jeopardizing the safety of the trunk's contents. In addition, the keys would work only if they were inserted right side up. The annoyance of such a system becomes acute when fumbling for the right key and right key alignment while trying to enter a locked car with a bag full of groceries, or a bundle of packages, which fall to the ground as a result of the fumbling.

In the mid-1970s, Toyota introduced a single key that unlocks the door, trunk, and glove compartment and starts the ignition. In addition the key could be inserted with either side up. This innovation was followed by other Japanese auto makers. However, more than 15 years later, G.M. has yet to adopt the idea and is not even considering it, now or in the near future. This *paradigm paralysis* exists despite research that shows **customers prefer** a single key. This example is not the only time G.M. has been caught in paradigm paralysis and resisted innovation. Other examples cited were resistance to seat belts, air bags, advanced engines, and transmissions. We believe that the principles and examples brought forth in this book will develop a *customer awareness* along with a set of problem-solving and decision-making skills that will develop creative responses to *customer needs*.

2.8 WORKING TOGETHER IN TEAMS

Even with a good road map, travelers may arrive at the wrong destination or take an excessive amount of time to get to the correct destination. In a similar way problem solvers may come to the wrong solution or take too much time to obtain a solution. Travelers also have to approach the trip with a positive attitude and draw upon the characteristics of expert travelers who have navigated the road before them. Finally, they need to make sure there is agreement on the route among themselves because conflicts between the travelers can make the trip unpleasant, sometimes with a disastrous outcome. Similarly, problem-solving groups can be much more efficient when they work well together. This chapter concludes with a section of ideas for enhancing group interactions to help the problem-solving process run smoothly.

As problems become more complex and interdisciplinary in nature, their solutions will require assembling groups of people with different areas of expertise. Consequently, it is important that the group dynamics of participants build cohesive, supportive, and productive teams. Because of the increasing focus in industry on the team approach to problem solving, many studies have been carried out on group interactions and how to improve them.[10-12]   Scholtes identifies ten common problems of team dynamics and some solutions to these problems.[9] Table 2-2 is an abbreviated summary.

Making
it
Work

TABLE 2-2: Top Ten List of Group Problems

Problem	How to minimize
1. Floundering	*Make sure the mission is clear and everyone understands what is needed to move forward.*
2. Overbearing Experts	*Have an agreement among team members that there are no sacred cows and that all team members have the right to explore all areas.*
3. Dominating Participants	*List "balance of participation" as a goal and evaluate regularly. Practice "gate keeping" to limit dominant participant.*
4. Reluctant Participants	*Ask opinions of quiet members and encourage by validation. Require individual assignments and reports.*
5. Unquestioned Acceptance of Opinion	*Ask for supporting data and reasoning. Accept and encourage conflicting ideas.*
6. Rush to Accomplishment	*Confront those doing the rushing and remind them not to compromise the best solution. Make sure a consensus is reached.*
7. Attribution of Motives to Others	*Reaffirm agreement that the group sticks to the scientific approach. Ask for confirmation of data.*
8. Discounting or Ignoring Group Member's Statement	*Provide training in effective listening. Support the discounted person. Talk off-line with any one who continually discounts other team members.*
9. Wanderlust: Digression and Tangents	*Follow an agenda with time estimates. Keep the topics in full view of the team and direct the conversation back to the topic.*
10. Feuding Team Members	*Focus on ideas, not personalities. Get adversaries to discuss the issues off-line or get them to agree to a standard of behavior during meetings.*

For a more complete discussion of ways to further minimize these top ten problems, the reader is referred to Scholtes.

Most problem-solving activities will require interaction with other people, either one-on-one or in group meetings. Meetings are essential tools for team problem solving. They should be carefully planned and skillfully run to realize the maximum benefit from them. The abbreviated list of guidelines shown in Table 2-3 will help team meetings be fruitful activities rather than time wasters![13,14]

TABLE 2-3: How to Run Effective Meetings

- At your first meeting, introduce yourselves, and give a little background. Set the group norms and expectations (e.g., showing up on time for meetings, responsibilities).

- Appoint a leader who will inspire the group to high levels of performance and be an effective listener.

- Prepare and distribute an agenda prior to meetings and stick to it.

- State why the group has come together.

- Bring all your materials (such as problem statement, group notes, handouts, your work, etc.) to all the meetings.

- Keep the discussion focused.

- Have someone in charge of keeping the meeting on track.

- Appoint a "devil's advocate" to challenge ideas as they arise.

- Have someone take minutes to remind participants of decisions made, actions to be taken.

- Draft an agenda for the next meeting, identifying what is to be covered and who is responsible for it.

Meeting Agenda
for 8/9/96

1. Open Meeting
2. Approve Minutes
3. Comments from
   the Chair
4. Committee Reports
5. Old Business
6. New Business
7. Set Next Meeting
8. Adjourn

The importance of meetings and positive group interactions cannot be overemphasized. For the problem-solving process to function smoothly, group members must get along. In many instances, the success of the project will depend upon how well people communicate and interact with one another.

CLOSURE

This chapter began by emphasizing the importance of approaching the problem with a positive "can do" attitude and striving to develop the traits of expert problem solvers. Next, we discussed the importance of taking risks as you formulate a truly innovative solution. We presented a heuristic which will be our guide or road map through the process of problem solving. The reader may wish to use the heuristic as presented or use it as a basis to develop his or her own heuristic or strategy. Finally, ways of fostering the creativity that can generate these innovative solutions and develop a positive atmosphere were presented, as were ideas for successfully interacting with colleagues as you develop your solution.

SUMMARY

- Practice
    The characteristics of expert problem solvers.
    The 7 Habits of Highly Effective People.
- Don't be afraid to take the risks necessary to obtain the very best solution.
- Welcome change and paradigm shifts as opportunities to make inroads and advancements.
- Look for voids in your organization and try to provide a vision to fill those voids to move the organization forward.
- Look for ways to use the road map of the five building blocks of the problem-solving heuristic.
- Develop an atmosphere that encourages and fosters creativity in those you work with.
- Listen to your *customers* and work *with* them as a unit to develop creative solutions.

REFERENCES

1. Whimbey, A., and J. Lochhead, *Problem Solving and Comprehension*: *A Short Course in Analytical Reasoning*, Franklin Press, Philadelphia, 1980.
2. Wankat, P.C., and F.S. Oreovicz, *Teaching Engineering*, McGraw-Hill, New York, 1993.
3. Matsen, J., *How to Fail Successfully*, Dynamo Publishing Co., Houston, TX, 1990.
4. *The Three Little Pigs*, Anonymous.
5. Barker, J.A., *Discovering the Future*, 2nd ed., ILI Press, St. Paul, MN, 1985.
6. Woods, D.R., *AIChE Symposium Series*, 79 (228), 1983.
7. Sculley, John, *Odyssey, Pepsi to Apple...A Journey of Advertising Ideas and the Future*, Harper & Row, New York, 1987.
8. Weldon, Joel, *Chemtech*, 13, p. 517, 1983.
9. Scholtes, Peter R., *The Team Handbook*, Joiner Associates, Inc., Madison, WI, 1988.
10. Bouton, C., and Y. C. Garth (eds.), *Learning in Groups, New Directions for Teaching and Learning*, 14, Jossey-Bass Inc., San Francisco, 1983.
11. Millis, B. "Helping Faculty Build Learning Communities through Cooperative Groups," *To Improve the Academy: Resources for Student, Faculty, and Institutional Devleopment*, 10, p. 43–58, 1990.
12. Berquist, W. H., and S. R. Phillips (eds.), *A Handbook for Faculty Development*, p. 118–121, Council of Independent Colleges, Washington, DC, 1975.
13. Patton, Robert R., *Solving Group Interaction*, Harper & Row, New York, 1973.
14. Shaw, Marvin E., *Group Dynamics*, McGraw-Hill, New York, 1976.

FURTHER READING

Gunneson, Alvin, "Communicating Up and Down the Parks," *Chemical Engineering*, 98, p. 135, June 1991. Useful tips on how to improve your interactions with those employees above, at the same level, and below you in your organization.

Raudsepp, Eugene, "Profits of the Effective Manager," *Chemical Engineering*, 85, p. 141, March 27, 1978. Although it was written 15 years ago, these traits still apply to effective leadership.

Phillips, Denise A., and A.E. Ladin Moore, "12 Commandments," *Chemtech*, 21, p. 138, March 1991. Rules to help improve your communication skills.

Strunk, W., and E. B. White, *The Elements of Style*, 3rd ed., Macmillan Publishing Co., New York, 1979. A concise treatise on grammar rules and writing style with many examples.

Whimbey, A., and J. Lockhead, *Problem Solving and Comprehension, A Short Course in Analytical Reasoning*, 2nd ed., The Franklin Institute Press, Philadelphia, 1980.

VanGundy Jr., Arthur B., *Techniques of Structured Problem Solving*, 2nd ed., Van Nostrand Reinhold, New York, 1988.

Lumsdaine, E., and M. Lumsdaine, *Creative Problem Solving, An Introductory Course for Engineering Students*, McGraw-Hill Publishing Co., New York, 1990.

EXERCISES

1. Make a list of the characteristics of expert problem solvers you would like to improve upon.

    A. _____

    _____

    B. _____

    _____

    C. _____

    _____

    D. _____

    _____

2. Choose three of the habits of highly effective people and explain how you will practice them during the coming weeks.

    A. _____

    _____

    B. _____

    _____

    C. _____

    _____

3. Describe three risks you can take during the coming months that will help to make you more comfortable with risk taking.

    A. _____

    _____

    B. _____

    _____

C. _____

_____

4. *Having a Vision.* Look around an organization of which you are a member for things that could be improved upon. Make a list of what is needed to make the organization an even better and more effective one.

A. _____

B. _____

C. _____

D. _____

E. _____

5. Which of the above changes would really alter the way the organization functions? What would need to be accomplished to produce a paradigm shift? How can you be a paradigm pioneer?

_____

_____

_____

_____

_____

_____

6. Identify a group of people with which you frequently interact. Make a list of things you can do to become a better team member and to establish a creative environment.

A. _____

B. _____

C. _____

D. _____

E. _____

7. Write a paragraph discussing your thoughts of Matsen's Failure 101 Class. Why is it important to learn how to contend with failure?

_____

_____

_____

_____

_____

_____

8. Collect one or more anecdotes (along with appropriate documentation/references) of initial failures that turned into major success stories (e.g., Domino's Pizza).

   _____

   _____

   _____

   _____

   _____

9. You are in a group of four working as a team to define and solve a problem. Describe how you would handle each of the following situations:
   a. Someone starts to dominate the group discussion and directions.
   b. Two of the group members are good friends and seem to form a clique.
   c. One of the group is not carrying their load.
   d. One of the group continually makes mistakes in their part of the project.

10. Prepare a list of specific ideas that would establish a creative environment in your group.
    A. _____
    B. _____
    C. _____
    D. _____
    E. _____

11. Develop an agenda for the first meeting of
    a. your colleagues to write a report for an undergraduate laboratory course.
    b. the floor of your residence hall to inform the new students of the rules, traditions, and other operations of the residence hall.
    c. a local interest group you are to lead.

12. *Working in Teams.* An instructor requested that students in the class form six-person teams, attempting to maximize diversity and selecting people to work with who were new to them. A team was formed that was composed of two white men, two white women, one African American man, and one Asian-American man. They selected one of the white men as their team leader. During the first and second meetings of this team, which took place during class time and in the classroom, the instructor noted that the African American man and the white man who was not the team leader sat almost outside the circle formed by the other team participants. Moreover, the white man who was the team leader, and one of the white women, appeared to do all the talking and to make all the suggestions about how to proceed. The other four people on the team looked uninvolved, at least as far as the instructor could observe. Questions: 1. What might be going on in this team? 2. Should the instructor intervene? Why? How? What would you do? 3. How might your own race/ethnicity and gender affect your options and choices about intervention? 4. What preparation, training, or instruction for teamwork might have helped this team? What training or instruction might be helpful to it now? 5. What preparation or instruction in teamwork dynamics, supervision, or intervention might be helpful to you in this and similar situations? (Adapted from the FAIRTEACH Workshop with the University of Michigan's School of Engineering Faculty, M. L. King, Jr., Day, 1994)

13. Match the icon to the proper habit of highly effective people.

A. _____

B. _____

C. _____

D. _____ 2 + 2 = 5

E. _____

F. _____

G. _____

3 PROBLEM DEFINITION

The mere formulation of a problem is far more often essential than its solution, which may be merely a matter of mathematical or experimental skill. To raise new questions, new possibilities, to regard old problems from a new angle requires creative imagination and marks real advances in science.

- Albert Einstein

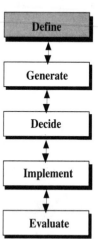

Often, the most difficult aspect of problem solving is understanding and *defining the real problem (sometimes also referred to as the underlying or root problem)*. In Chapter 1, we presented a number of true examples of incorrectly defined problem statements that demonstrate how competent, conscientious people can define the wrong problem and waste considerable time looking in the wrong direction for a solution. In this chapter we address the first part of the heuristic, *problem definition*. A study that we conducted of experienced problem solvers in industry revealed some common threads that run through their problem definition techniques. We have classified these common threads into a number of steps that can help you understand and define the real problem.

3.1 THE FIRST FOUR STEPS

The first four steps used by experienced problem solvers to understand and define the real problem are given in Table 3-1. You will observe that the first four steps focus on gathering information.

> ### TABLE 3-1: What Experienced Problem Solvers Say
> 1. Collect and analyze information and data.
> 2. Talk with people familiar with the problem.
> 3. If at all possible, view the problem first hand.
> 4. Confirm all findings.

Step 1. *Collect and analyze information and data.*

Learn as much as you can about the problem. Write down or list everything you can think of to describe the problem. Until the problem is well defined, anything might be important. Determine what information is missing and what information is extraneous. The information should be properly organized, analyzed and presented. It will then serve as the basis for subsequent decision making. Make a simple sketch or drawing of the situation. Drawings, sketches, graphs of data, etc. can all be excellent communication tools when used correctly. Analyze the data to show trends, errors, and other meaningful information. Display numerical or quantitative

"Start with an open mind."

"Don't jump to conclusions."

"Look at the big picture."

"Review the obvious."

data graphically rather than in tabular form. Tables can be difficult to interpret and sometimes misleading. Graphing, on the other hand, is an excellent way to organize and analyze large amounts of data. Methods for plotting data to reveal trends are given in Appendix 2. The Case of the Dead Fish provides an interesting example of the use of graphical data to solve problems.

The Case of the Dead Fish

Research and information gathering are great tools in problem solving. We consider the case of a chemical plant that discharges waste into a stream that flows into a relatively wide river. Biologists monitored the river as an ecosystem and reported the following data of the number of dead fish in the river and the river level:

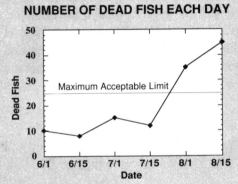

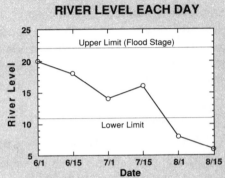

Graphs of the type shown above are called time plots and control charts. A time plot shows trends over a period of time (e.g., the level of a river over several days or weeks). A control chart is a time plot that also shows the acceptable limits of the quantity being displayed. For example, in the control chart of the river level, the upper and lower acceptable water levels would also be shown. If one of the acceptable limits is exceeded, this occurrence may yield some information about the timing of the problem and possible causes of it. We can then examine time plots of other pertinent quantities and look for additional clues about the problem,

From the graphs we see that the acceptable level of dead fish was exceeded on August 1 and 15. We look for anything that might have occurred on or between July 15th and August 1. We discover that on July 29 there was a large amount of chemical waste discharged into the river. Discharges of this size had not caused any problems in the past. Upon checking other factors, we see that there has been little rain and that the water level in the river, measured on August 1, had fallen so low it might not have been able to dilute the plant's chemical waste. Consequently, the low water level, coupled with the high volume of waste, could be suggested as a possible cause for the unusually large number of dead fish. However, to verify this, we would have to carry the analysis further. Specifically, we shall soon use one or more of the problem definition techniques discussed later in this chapter.

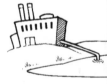

Step 2. *Talk with people familiar with the problem.*

Find out who knows about the problem. Ask penetrating questions by

- Looking past the obvious
- Challenging the basic premise
- Asking for clarification when you do not understand something

Ask *Insightful* Questions

Our experience shows that seemingly naive questions (often perceived as "dumb" questions) can produce profound results by challenging established thinking patterns. This act of *challenging* must be an ongoing process.

You should also talk to other people about the problem. Verbalizing the problem to someone else helps clarify in your own mind just what it is you are trying to do. Try to find out who the experts in the field are and talk to them. Nonexperts are also a rich source of creative solutions, as evidenced by the following example.

Seeking Advice

Joel Weldon, in his tape "Jet Pilots Don't Use Rearview Mirrors," described a problem encountered by a major hotel a number of years ago. Since the hotel had become very popular, the elevators were very busy, and frequently caused backups in the lobby area. The manager and assistant manager were lamenting the problem in the lobby one day and were brainstorming about how to increase the elevator capacity. Adding additional elevator shafts would require removal of a number of rooms and a significant loss of income. The doorman, overhearing their conversation, casually mentioned that it was too bad they couldn't just add an elevator on the outside of the building, so as not to disturb things inside. A great idea! It occurred to the doorman because he was outside the building much of the time, and that was his frame of reference. Notice, however, that the doorman's creativity alone was not enough to solve the problem. Knowledge of design techniques was necessary to implement his original idea. A new outside elevator was born, and the rest is history. External elevators have since become quite popular in major hotels. Information, good ideas, and different perspectives on the problem can come from all levels of the organization. (*Chemtech.*, 13, 9, p. 517, 1983)

When equipment malfunctions, it is a *must* to talk to the operators because they know the "personality" of the equipment better than anyone.

Most organizations have employees who have "been around a long time" and have a great deal of experience, as illustrated in the following example.

Go Talk to George

Remember the leaking flowmeter discussed in Chapter 1? The solution that the company adopted was to replace the flowmeter at regular intervals. Let's consider a similar situation in which, immediately upon replacement, the flowmeter began to leak. List in order four people you would talk to.

- the person who installed the meter
- the technician who monitors the flowmeter
- the manufacturer's representative who sold you the flowmeter
- George

Who's George? Every organization has a *George*. George is that individual who has years of experience to draw upon and also has street smarts. George is an excellent problem solver who always seems to approach the problem from a different viewpoint– one that hasn't been thought of by anyone else. Be sure to tap this rich source of knowledge, when you are faced with a problem. Individuals such as George can often provide a unique perspective on the situation.

Step 3. *View the problem first hand.*

> "You can see a lot just by looking."
> –Yogi Berra

While it is important to talk to people as a way to understand the problem, you should not rely solely on their interpretations of the situation and problem. If at all possible, go inspect the problem yourself.

Viewing the Problem Firsthand

In the mid 1970s a company in the United Kingdom completed a plant to produce a plastic product (PVC). The main piece of equipment was a large reactor with a cooling jacket through which water passed to keep the reactor cool. When the plant was started up, the plastic was dark, nonuniform, and way off design specifications. The engineers in charge reviewed their design. They reworked and refined their model and calculations. They analyzed the procedure from every point of view on paper. They had the raw material fed to the reactor analyzed. However, they all came up with the same results– that the product should definitely meet the design specifications. Unfortunately, nobody examined the reactor firsthand. Finally after many days, one of the engineers decided to look into the reactor. He found that a valve had been carelessly switched to the wrong position, thereby diverting cooling water away from the reactor so that virtually no cooling took place. As a result the reactor overheated, producing a poor quality product. Once the valves were adjusted properly, a high quality plastic was produced.

- continued -

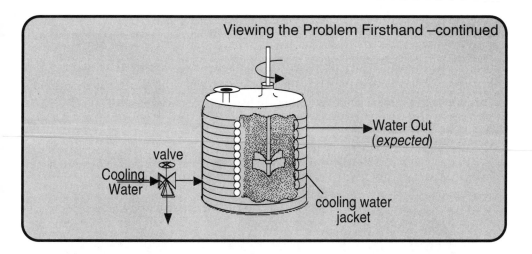

Viewing the Problem Firsthand –continued

Water Out
(*expected*)

valve

Cooling
Water

cooling water
jacket

Step 4. *Confirm all key findings.*

Verify that the information that you collected is correct. Cross check and cross reference data, facts, and figures. Search for biases or misrepresentation of facts. Confirm all important pieces of information and spot check others. Distinguish between fact and opinion. Challenge assumptions and assertions.

Confirm All Allegations

The authors of this book were involved in a consulting project for a pulp and paper company we will call Boxright. Several years ago, Boxright had installed a new process for recovering and recycling their "cooking" chemicals used in the papermaking process. Two years after the installation, the process had yet to operate correctly. Tempers flared and accusations flew back and forth between Boxright and Courtland Construction, the supplier of the recycling equipment. Courtland claimed the problem was that Boxright did not know how to operate the process correctly, while the company contended that the equipment was improperly designed. Boxright finally decided to sue Courtland for breach of equipment performance. Much data and information were presented by both sides to support their arguments. Courtland presented data and information from an article in the engineering literature that they claimed *proved* Boxright was not operating the process correctly. At this point it looked like Courtland had cooked our goose by presenting such data. However, before conceding the case we needed to confirm this claim. We analyzed this key information in detail, and to our glee found in the last few pages of the article it was stated that the data would not be expected to apply to industrial-size equipment or processes. When this information was presented, the lawsuit was settled in favor of the pulp and paper company, Boxright.

3.2 DEFINING THE REAL PROBLEM

The four steps just discussed are all related to gathering information about the problem. This information lays the groundwork that will help us use the problem definition techniques discussed in this chapter.

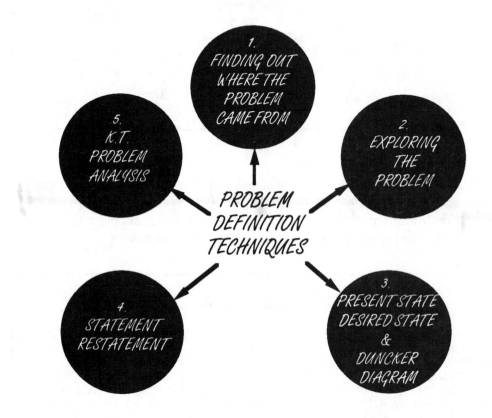

These techniques are used to help understand the problem so that we may define the real problem as opposed to the perceived problem. The K.T. Problem Analysis will be discussed in Chapter 5 as part of the Kepner-Tregoe Approach.[1]

3.2A Finding Out Where the Problem Came From

Many times you will be given a problem by someone else rather than discovering it yourself. Under these circumstances, it is very important that you make sure that the problem you were given reflects the true situation. This technique focuses on finding out who initiated the problem and ascertaining the validity of the reasoning used to arrive at the problem statement.

Find out where the problem statement came from.

- Where did the problem originate?

- Who posed the problem statement in the first place, your supervisor, his/her supervisor, a colleague in your project group, or someone else?

- Can that person explain the reasoning as to how they arrived at that particular problem statement?

- Are the reasoning and assumptions valid?

- Has that person considered the situation from a number of different viewpoints before arriving at the final problem statement?

- Have you used *the first four steps* to gather information about the problem?

Try to detect any errors in logic as you trace the problem back to its origins. Distinguish opinion from fact and conclusions from evidence. **Never** assume that the problem statement you were given is correctly worded or has been thoroughly investigated.

> Trace the problem back to its origin.

Always check to be sure that the problem statement directs the solution to the true cause and does not seek merely to treat the symptoms. For example, it would certainly be better to find the cause of the off-taste in the hamburgers at a fast food chain rather than treating the symptoms by adding more spices to cover the off-taste. Make certain that time and energy are not wasted merely dealing with the symptoms.

Remember *The Case of the Dead Fish* in the river on p.30? The dead fish example is a case where giving directions to treat the symptoms rather than discovering the real cause of the problem could have lead to a costly, unnecessary solution.

Finding Out Where the Problem Came From
The Case of the Dead Fish

The Situation: Stan Wilson is an engineer with six years of experience with his company. The instructions given by Stan's supervisor to solve the perceived problem: *"Design a new waste treatment plant to reduce the toxic waste from the chemical plant."* Stan and his team are requested to design treatment facilities to reduce the toxic chemical concentrations by a factor of 10. A quick back-of-the-envelope calculation shows that the plant could cost well over a *million* dollars. Stan is puzzled because the concentrations of toxic chemicals have always been significantly below governmental regulations and company health specifications.

Who posed the problem?

Stan approaches his supervisor to learn more about the reasons for the order. The supervisor informs Stan that it was not his decision, but upper management's.

Can reasons for arriving at the problem statement be explained?

The supervisor tells Stan it has something to do with the summer drought and a number of recent articles in the local newspaper about the unusually high number of dead fish that have turned up in the river in the last few weeks. He said that it was his understanding that the drought has brought the river to an extremely low level and that the discharge was no longer sufficiently dilute to be safe to the fish and other aquatic life. Consequently, to deflect the negative press and avoid possible lawsuits, the company has announced the planning of a new waste treatment facility.

Are the assumptions and reasoning valid?

> Keep digging to learn the motivation (who, why) for issuing the instructions to solve the perceived problem.

Thus, Stan realizes that the decision to design and build a waste treatment plant is based on *an unusually large number of dead fish in the river, and **not** necessarily on the presence of high concentrations of toxic chemicals*. The company had decided to try to treat the symptoms (many dead fish) by removing toxic chemicals, thus solving the perceived problem, but not necessarily the true problem (how to prevent the fish from dying).

Has sufficient data/information been collected?

In the Explore Phase, we'll see how Stan initiated his own investigation into the case of the dead fish and eventually found the true cause of the problem.

Finding Out Where the Problem Came From
Sweet and Sour

The Situation: (which has nothing to do with Chinese food):   Natural gas (methane), which contains significant levels of hydrogen sulfide, is called a *sour gas*, while natural gas that does not contain hydrogen sulfide is called a *sweet gas*. Sour gas is particularly troublesome because it is extremely corrosive to the pipes and equipment used to transport it. Tom Anderson was the sour gas piping expert at a major oil company that was drilling an off-shore well in a gas field in the North Sea. Regions near the well being drilled were known to produce sour gas. Tom received a call from the head office. <u>The instructions given to solve the perceived problem</u>: *"Fly to Copenhagen to begin the design and installation of a piping system that would transport sour gas from the new well to the platform facility."*

Laboratory tests were believed to have been carried out on gas samples from this well and it was assumed that the head office had reviewed these tests. An expensive piping system that would be resistant to corrosion by sour gas was designed and installed. When the gas well was brought on-line, it was found that the gas was sweet gas which did not require the corrosion resistant piping system that had cost several million dollars extra.

Who was responsible for this blunder? Could this waste have been eliminated if Tom had found out where the problem had come from? Did the problem come from the lab or from the head office? What would have been the course of action regarding the type of piping installed,

1. **If** Tom, who was the piping expert, had asked the head office to explain why they wanted to install piping resistant to sour gas for *this* well, or,
2. **If** Tom had challenged their reasoning by asking what evidence they had that *this* well produced sour gas, or
3. **If** Tom had gathered more information by tracking down the laboratory results to learn how much sour gas was in the natural gas?

If Tom had traced back the original source of the product to find out **where the problem came from**, this waste could have been eliminated.

> Challenge
> Assumptions
> and
> Reasoning

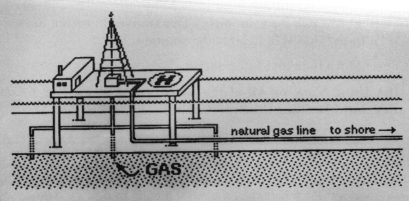

A good rule of thumb is to treat the symptoms *only* if it is impossible or impractical to solve the real problem. For example, in the case of the leaking flowmeter discussed in Chapter 1, it was impractical to solve the real problem of finding an inexpensive corrosion resistant material. As a result, the symptoms were treated by periodically replacing the flowmeter.

3.2B Exploring the Problem

This technique works well both for situations of analyzing incorrectly defined problems assigned to you and for formulating problem statements for new problems you uncover yourself. Once presented with a problem, we want to explore all aspects of the problem and its surroundings. This technique, which has its origins in the McMaster Five-Point Strategy given in Appendix 1, is a procedure that guides us to understand and define the real problem. Gathering information is also the key to the success of the exploration, and *the first four steps* (p. 29) are very helpful in this process.

TABLE 3-2: Exploring the Problem[2]

1. **Identify All Available Information.**

2. **Recall or Learn Pertinent Theories and Fundamentals.**

3. **Collect Missing Information.**

4. **Solve a Simplified Version of the Problem to Obtain a "Ballpark" Answer.**

5. **Hypothesize and Visualize What Could Be Wrong with the Current Situation.**

6. **Brainstorm to Guess the Answer.**

7. **Recall Past or Related Problems and Experiences.**

8. **Describe or Sketch the Solution in a Qualitative Manner or Sketch Out a Pathway That Will Lead to the Solution.**

9. **Collect More Data and Information.**

10. **After Using Some or All of the Activities Above, Write a Concise Statement Defining the Real Problem.**

"Exploring the Problem" can also be used to build upon the results of the previous technique "Finding Out Where the Problem Came From."

Exploring the Problem

The Case of the Dead Fish

Stan decides to initiate his own investigation into the dead fish problem over the weekend.

1. **Identify Available Information**: There is a toxic discharge from the plant, the river level is low, and there are a large number of dead fish in the river.

2. **Learn Fundamentals**: Stan calls a friend in the biology department at the local university and asks her about the problem of what could be causing the fish to die. She tells Stan that the extremely low water levels lead to significantly warmer water temperatures, and hence lower levels of dissolved oxygen in the water. These conditions make the fish susceptible to disease.

3. **Missing Information**: Secondly, she says that a fungus has been found in two nearby lakes that could be responsible for the death of the fish. Upon checking the recent daily temperatures, Stan learns that the day before the fish began dying was one of the hottest of the decade. Stan starts making phone calls to people upstream and downstream from the plant and learns that dead fish are appearing at the same unusually high rate everywhere, not just downstream of the plant.

5. **Hypothesis**: The fish were dying all over the area as a result of the fungus, and not from the plant discharge.

9. **More Information**: Upon examination of the dead fish, it was discovered that the fungus was indeed the cause of death, and that toxic chemicals played no role in the problem. Stan was glad that he had found out where the problem had come from and had explored the situation rather than blindly proceeding to design the treatment plant.

10. **Define the Real Problem**: Identify ways to cure the infected fish and prevent healthy fish from being infected.

We note from the above example that it is not always necessary to address all ten steps in Table 3-2 to fully explore the problem. However, as seen in the next example, each of the steps has a purpose and contributes to revealing the true problem.

De-bottlenecking a Process

Even though the following example is taken from an actual case history, don't worry if you don't know much about heat exchangers; just follow the reasoning. It is too good an example to pass up. <u>The situation</u>: A valuable product was being sold as fast as it could be manufactured in a chemical plant. Management tried to increase production but was unable to do so. Analysis of each step in the production line showed that the bottleneck was the refrigeration unit. This unit was a simple heat exchanger in which the hot liquid stream was cooled by passing it through a pipe which contacted a cold liquid stream. Heat flowed from the hot stream through the pipe wall into the cold stream. Unfortunately the refrigeration unit (i.e., heat exchanger) was not cooling the hot liquid stream to a sufficiently low temperature for it to be treated effectively in the next processing step. <u>The instructions given to solve the perceived problem</u>: *"Design and install a larger refrigeration unit."* The design of a larger refrigeration unit was started.

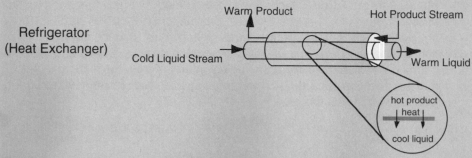

Refrigerator (Heat Exchanger)

Explore Phase

1. **Identify inputs/outputs:** Cold liquid stream not cooling hot product stream.

2. **Recall related theories and fundamentals:** The rate of cooling between the two streams is related to the temperature difference between the two streams, their flow rates, and the materials and condition of the unit.

3. **Collect missing information:** What is the size of the current refrigeration unit? What are the entering and exiting temperatures of the liquid streams?

4. **Carry out an order of magnitude calculation:** AH HA! The new unit need be no larger than the old one.

5. **Hypothesize and visualize what could be wrong with the current system:** Inefficient operation of current system? Could something be increasing the resistance to heat transfer (i.e., insulating)?

6. **Guess the result:** Could scale (minerals deposited from the liquid) have built up on the inside of the unit acting as an insulating blanket?

- continued -

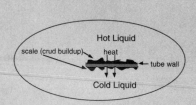

De-bottlenecking a Process –continued

Hot Liquid

scale (crud buildup)   heat

◄— tube wall

Cold Liquid

The buildup of scale on the pipe walls of the exchanger reduces the amount of heat that will transfer from the hot fluid to the cold fluid which severely degrades the ability of the exchanger to perform its intended task. The thicker the scale, the greater the resistance to heat transfer and the poorer the performance of the unit.

7. **Recall past problems, theories, or related experiences:** Scale greatly reduces the efficiency of the unit.

8. **Sketch solution or solution pathway:** Examine the unit for evidence of scale or fouling that may be reducing the heat transfer efficiency.

9. **Collect more data:** An examination of the heat exchanger showed it was indeed badly fouled.

10. **Define the real problem:** The scale on the pipe wall must be removed in order to cool the product stream effectively.

3.2C Using the Present State/Desired State Technique

How many times have you heard the statement "You can't get there from here?" The *Present State/Desired State* technique helps us verbalize where we are and where we want to go, so that an appropriate path can be found and we can indeed get there from here. The Present State/Desired State technique also helps us learn whether the solution goals (Desired State) are consistent with our needs (Present State).[3] When writing the Desired State statement, avoid using ambiguous and vague words or phrases like "best," "minimal," "cheapest," "within a reasonable time," "most efficient," etc. because these words mean different things to different people. Be quantitative where possible. For example, "The children's playground needs to be completed by July 1, 1994 at a cost under $100,000" *as opposed to* "The playground should be completed in a reasonable time at minimal cost." It is important that the Present State statement match the Desired State statement. In order for the Present State and Desired State to match, every concern in the Present State should be addressed in the Desired State. In addition, the Desired State should not contain solutions to problems that are not in the Present State. Sometimes a match exists, but it really doesn't get to the heart of the problem or allow many solution alternatives. Reworking the Present State and Desired State statements until they match is a technique that increases the probability of arriving at the true problem statement. Let us consider the following example of the Present State/Desired State Technique.

Cleaning up the
Problem Statement

Hitting 'Em Where They Aren't

<u>The Situation</u>: During WWII, a number of aircraft were shot down while engaging in bombing missions over Germany. Many of the planes that made it back safely to base were riddled with bullet and projectile holes. The damaged areas were similar on each plane.

<u>The instructions given to solve the perceived problem</u>: *"Reinforce these damaged areas with thicker armor plating."*

(Present State)   **(Desired State)**

Many bullets/projectiles penetrating aircraft.   Fewer planes being shot down.

<u>*Discussion*</u>: This is not a match because there are planes that are surviving that still have bullet holes.   There is not a *one-to-one mapping* of all the needs of the present state being addressed and resolved in the desired state.

(Present State)   **(Desired State)**

Many bullets/projectiles penetrating aircraft.   Fewer bullet holes.

<u>*Discussion*</u>: These states are matched, but the distinction between the present state and the desired state is not clear enough. It may take only a single bullet hitting a critical area to down a plane.

(Present State)   **(Desired State)**

Many bullets/projectiles penetrating aircraft   Fewer bullets/projectiles
in critical and noncritical areas.   penetrating critical areas.

<u>*Discussion*</u>: These two statements now match and the distinction between them is sharp, opening up a variety of solution avenues such as reinforcing critical areas, moving critical components (e.g., steering mechanism) to more protected locations, providing redundant critical components, etc.

Note: The original instructions given to solve the perceived problem would have failed. Reinforcing the areas where returning planes had been shot would have been futile. Clearly these were noncritical areas; otherwise these planes would have been casualties as well.

3.2D The Duncker Diagram

The Duncker Diagram helps obtain solutions that satisfy the criteria set up by the Present State/Desired State statements.[3] The unique feature of the Duncker Diagram is that it points out ways to solve the problem by making it OK *not* to reach the desired solution. Duncker Diagram solutions can be classified as General Solutions, Functional Solutions, and Specific Solutions (see Figure 3-1).

There are two types of General Solutions: 1) Solutions on the left side of the diagram that move from the present state to the desired state (i.e., we have to do something) and 2) solutions on the right side that show how to modify the desired state until it corresponds with the present state (make it OK *not* to do that *same* something). For example, suppose your *present state* was your current job and the *desired state* is a new job. The left hand side of the diagram would show the steps to reach the desired state of obtaining a new job (e.g., update resume, interview trips). The right side of the diagram show the steps that would make it OK to stay in your current job (e.g., greater participation in the decision making, salary increase). In addition, there could be a compromise solution in which both the Present State and Desired State are moved toward each other until there is a correspondence.

     Functional Solutions are possible paths to the desired state (or modified desired state) that do not take into account the feasibility of the solution. We could solve the problem *only if...* we had more time, more personnel, ... we won the lottery.... After arriving at each functional solution, one has to suggest feasible *Specific Solutions* to implement the functional solutions. For example in the job change situation, a functional solution on the right side of the diagram might be feeling more appreciated and a specific solution to feeling appreciated could be a salary increase or bonus, more verbal praise on a job well done, or a letter of commendation in your company personnel file. Representing the problem on a Duncker Diagram is a creative activity, and as such, there is no right way or wrong way to do it. There are only more and less useful ways to represent the problem. Typically, the most difficult activity is choosing the appropriate desired state. This skill improves with practice.

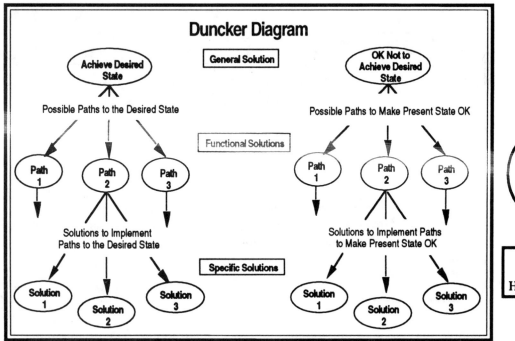

Figure 3-1. The Duncker Diagram

Kindergarten Cop[†]

Linda Chen, who has been teaching elementary school for 25 years, has just finished a six-month leave of absence and is scheduled to return to teaching in February. She is dreading returning to teaching because the last few years have been extremely stressful and difficult, and she feels burned out teaching kindergarten. Students seem harder to control, Linda doesn't like the materials she is required to use in the classroom, and the parents don't seem to take much interest in their children's education. She also enjoyed the time she had to herself during her six-month leave and strongly feels she must continue to have more time to herself as she nears retirement which will be in five years if she is to receive full benefits. Consequently, Linda's *present state* is returning to teaching, and her *desired state* is not to return to teaching. Prepare a Duncker Diagram to analyze this situation.

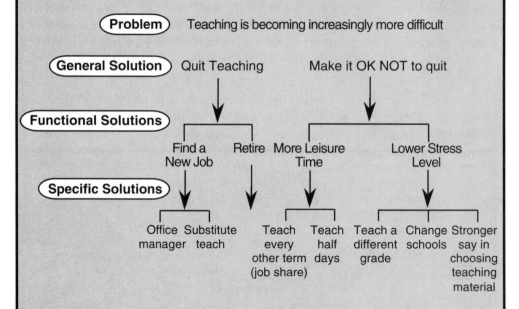

(Problem)   Teaching is becoming increasingly more difficult

(General Solution)   Quit Teaching   Make it OK NOT to quit

(Functional Solutions)

     Find a   Retire   More Leisure   Lower Stress
     New Job   Time   Level

(Specific Solutions)

| Office manager | Substitute teach | Teach every other term (job share) | Teach half days | Teach a different grade | Change schools | Stronger say in choosing teaching material |

Recap: Upon analyzing her situation using a Duncker Diagram, Linda discovered the real problem was the high stress level brought on by the unruly classes she had the year before her six-month leave. Consequently, with the aid of a Duncker Diagram, she arrived at the conclusion that the *real* problem was she should find ways to lower her stress level at her workplace.

[†]Based on an actual case history.

Let us consider the application of the Duncker Diagram to the following To Market, To Market example.

To Market, To Market

The Situation: Toasty O's was one of the hottest selling cereals when it first came on the market. However, after several months, sales dropped. The consumer survey department was able to identify that customer dissatisfaction was expressed in terms of a stale taste. <u>The instructions given to solve the perceived problem</u>: *"Streamline the production process to get the cereal on the store shelves faster, thus ensuring a fresher product."* However, there wasn't much slack time that could be removed from the process to accomplish the goal. Of the steps required to get the product on the shelves (production, packaging, storage, and shipping), production was one of the fastest. Thus, plans for building plants closer to the major markets were considered, as were plans for adding more trucks in order to get the cereal to market faster. The addition of new plants and trucks was going to require a major capital investment to solve the problem.

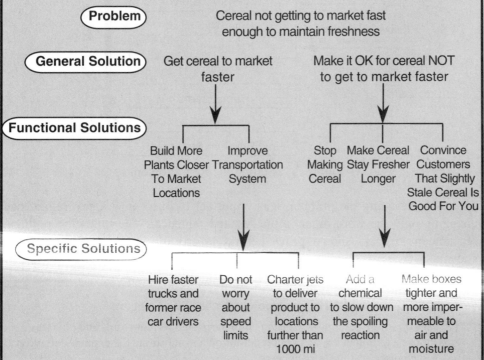

Problem   Cereal not getting to market fast enough to maintain freshness

General Solution   Get cereal to market faster   Make it OK for cereal NOT to get to market faster

Functional Solutions

Build More Plants Closer To Market Locations   Improve Transportation System   Stop Making Cereal   Make Cereal Stay Fresher Longer   Convince Customers That Slightly Stale Cereal Is Good For You

Specific Solutions

Hire faster trucks and former race car drivers   Do not worry about speed limits   Charter jets to deliver product to locations further than 1000 mi   Add a chemical to slow down the spoiling reaction   Make boxes tighter and more impermeable to air and moisture

The **real problem** was that the cereal was not staying fresh long enough, not that it wasn't getting to market fast enough. Keeping the cereal fresher longer was achieved by improved packaging and the use of additives to slow the rate of staling.

3.2E Using the Statement-Restatement Technique

> A problem well-stated is a problem half solved.
> -Charles F. Kettering

This technique is similar to the Present State/Desired State technique in that it requires us to rephrase the problem statement. The *Statement-Restatement* technique was developed by Parnes,[4] a researcher in problem solving and creativity. Here, one looks at the fuzzy or unclear problem situation and writes a statement regarding a challenge to be addressed. The problem is then restated in different forms a number of times. Each time the problem is restated, one tries to generalize it further in order to arrive at the broadest form of the problem statement (see Figure 3-2).

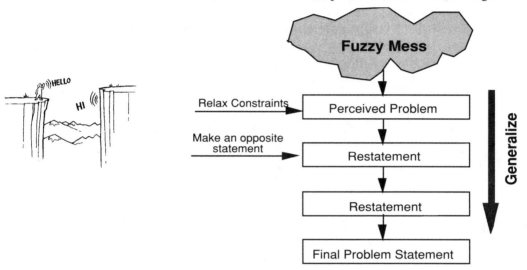

Figure 3-2. Stating the Real Problem

In restating the problem it is important to inject new ideas, rather than changing only the word order in the restated sentence. The following problem restatement *triggers* should prove helpful in arriving at a definitive problem statement.

TABLE 3-3: Problem Statement Triggers

1. Vary the stress pattern—try placing emphasis on different words and phrases.
2. Choose a term that has an explicit definition and substitute the explicit definition in each place that the term appears.
3. Make an opposite statement, change positives to negatives, and vice versa.
4. Change "every" to "some," "always" to "sometimes," "sometimes" to "never," and vice versa.
5. Replace "persuasive words" in the problem statement such as "obviously," "clearly," and "certainly" with the argument it is supposed to be replacing.
6. Express words in the form of an equation or picture, and vice versa.

Using the Triggers

Original Problem Statement: Cereal not getting to market fast enough to maintain freshness

Trigger 1
- <u>Cereal</u> not getting to market fast enough to maintain freshness. (Do other products we have get there faster?)
- Cereal not <u>getting</u> to market fast enough to maintain freshness. (Can we make the distance/time shorter?)
- Cereal not getting to <u>market</u> fast enough to maintain freshness. (Can we distribute from a centralized location?)
- Cereal not getting to market fast enough to maintain <u>freshness</u>. (How can we keep cereal fresher, longer?)

Trigger 2
- <u>Breakfast food that comes in a box</u> is not getting to <u>the place where it is sold</u> fast enough to keep it from <u>getting stale.</u> (Makes us think about the box and staleness. . . what changes might we make to the box to prevent staleness?)

Trigger 3
- How can we find a way to get the cereal to market <u>so slowly</u> that it will <u>never</u> be fresh? (Makes us think about how long we have to maintain freshness and what controls it?)

Trigger 4
- Cereal is not getting to market fast enough to <u>always</u> maintain freshness. (This change opens new avenues of thought. Why isn't our cereal <u>always</u> fresh?)

Trigger 5
- The problem statement implies that we obviously want to get the cereal to market faster to maintain freshness. Thus, if we could speed up delivery freshness would be maintained. Maybe not! Maybe the store holds it too long. Maybe it's stale before it gets to the store. (This trigger helps us challenge implicit assumptions made in the problem statement.)

Trigger 6
- Freshness is inversely proportional to the time since the cereal was baked, i.e.,

$$(\text{Freshness}) = \frac{k}{(\text{Time Since Baked})}$$

Makes us think of other ways to attack the freshness problem. For example, what does the proportionality constant, k, depend upon?

The storage conditions, packaging, type of cereal, etc. are logical variables to examine. How can we change the value of k?

The total time may be shortened by reducing the time at the factory, the delivery time, or the time to sell the cereal (i.e., shelf time). So, again, this trigger provides us with several alternative approaches to examine to solve the problem: Reduce the time <u>or</u> change (increase) k.

As an illustration of the use of these triggers, consider *trigger 3* above. Instead of asking "How can my company make the biggest profit?" ask "How can my company lose the most money?" In finding the key activities or pieces of equipment which, when operated inefficiently, will give the biggest loss, we will have found those pieces that need to be carefully monitored and controlled. This trigger helps us find the *sensitivity* of the system and to focus on those variables that dominate.

It is often helpful to *relax constraints* on the problem, modify the criteria, and idealize the problem when writing the restatement sentence (see trigger 4). Also, does the problem statement change when different time scales are imposed (i.e., are the long-term implications different from the short-term implications)? As one continues to restate and perhaps combine previous restatements, one should also focus on tightening up the problem statement, eliminating ambiguous words, and moving away from a fuzzy, loose, ill-defined statement.

Making an Opposite Statement

The Situation: To many people, taking aspirin tablets is a foul-tasting experience. A few years ago, a number of companies making aspirin decided to do something about it. <u>The instructions given by the manager to his staff to solve the perceived problem were</u>: *"Find a way to put a pleasant-tasting coating on aspirin tablets."* Spraying the coating on the tablets had been tried, with very little success. The resulting coating was very nonuniform and this led to an unacceptable product. Let's apply the triggers to this problem.

Trigger 1   Emphasize different parts of statement
    1. **Put** coating **on** tablet.

Trigger 3   Make an opposite statement
    2. **Take** coating **off** tablet.

This idea led to one of the newer techniques for coating pills. The pills are immersed in a liquid which is passed onto a spinning disk. The centrifugal force on the fluid and the pills causes the two to separate, leaving a nice, even coating around the pill.

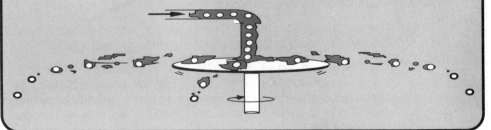

An example from reliable, although undocumented, sources that elucidates the need to find the real problem is one related to the early research on the reentry of space capsules to the earth's atmosphere. It was evident that available materials would not withstand the temperatures from frictional heating by the atmosphere. Consequently, a directive went out to find a material able to withstand the temperatures encountered on reentry. Application of the Statement-Restatement technique to this problem is shown in the following gray box.

The real problem here was to protect the astronauts (restatement 3) rather than find a material that would withstand high temperatures. Once the real problem was found, an appropriate solution to the capsule reentry problem soon followed.

Sacrificial
Nosecone

Wanted: Exotic Materials, or...?

The Situation: In the 1960s scientists recognized that there was no available material that would survive the high temperatures generated on the capsule's surface during reentry to the earth's atmosphere. <u>Consequently, a government directive went out</u> to *"find a material able to withstand the temperatures encountered on reentry."* By the early 1970s no one had produced a suitable material that satisfied the directive, yet we had sent astronauts to the moon and back. How had this achievement been possible? The **real problem** was to protect the astronauts upon reentry, rather than to find a material that would withstand such high temperatures. Once the real problem was determined, a solution soon followed. One of the scientists working on the project asked a related question: How do meteors eventually reach the earth's surface without disintegrating completely? Upon investigation of this problem, he found that although the surface of the meteor vaporized while passing through the atmosphere, the inside of the meteor was not damaged. This analogy led to the idea of using materials on the outside of the capsule that would vaporize when exposed to the high temperatures encountered during reentry. Consequently, the heat generated by friction with the earth's atmosphere during reentry would be dissipated by the vaporization of a material that coated the outside of the space capsule. By sacrificing this material, the temperatures of the capsule's underlying structural material remained at a tolerable level to protect the astronauts. Once the real problem was uncovered, the scientists solved the problem by using analogies and transferring ideas from one situation to another.

Statement-Restatement

The statement-restatement technique might have been used as follows:

Statement 1: Find a material that will withstand the high surface temperature of the capsule resulting from frictional heating upon reentry into the earth's atmosphere.

Restatement 1: Find a way to slow the reentry into the earth's atmosphere or to redesign the capsule so that the capsule surface temperature will be lower.

Restatement 2: Find a way to cool the capsule or absorb the frictional energy during reentry so that the surface temperature will be lower.

Restatement 3 Find a way to protect the astronauts on their reentry into the earth's atmosphere.

Restatement 4: Find a disposable material that could surround the capsule and could be sacrificed to absorb the frictional heating.

3.2F Evaluating the Problem Definition

Now that we have used one or more of the preceding techniques to define the problem, we need to check to make sure we are going in the right direction. Consequently, we need to evaluate the problem definition before proceeding further. The following checklist could help us in this evaluation.

> If you break the problem apart . . . be sure to put it back together.

- Have all the pieces of the problem been identified?
- Have all the constraints been identified?
- What is missing from the problem definition?
- What is extraneous to the problem definition?
- Have you challenged the assumptions and information you were given to formulate the problem?
- Have you distinguished fact from opinion?

3.3 THE NEXT FOUR STEPS

We now extend the first steps experienced problem solvers recommend and continue Table 3-1 in Table 3-4.

TABLE 3-4: What Experienced Problem Solvers Say

The First Four Steps of Experienced Problem Solvers

1. Collect and analyze information and data.
2. Talk with people familiar with the problem.
3. If at all possible, view the problem firsthand.
4. Confirm all findings.

The Next Four Steps

5. Determine if the problem should be solved.
6. Continue to gather information and search the literature.
7. Form simple hypotheses and quickly test them.
8. Brainstorm potential causes and solution alternatives.

Step 5. *Determine if the problem should be solved.*

> Establish criteria to Judge the Solution.

Having defined the real problem, we now need to develop criteria by which to judge the solution to the real problem. One of the first questions experienced engineers ask is: Should the problem be solved? Figure 3-3 shows how to proceed to answer this question. The first step is to determine if a solution to an identical or similar problem is available. A literature search may determine if a solution exists.

How do experienced problem solvers go about deciding if the problem is <u>worth</u> solving? Perhaps it is just mildly irritating and consequently may be ignored altogether. (For instance, suppose the garage door at your plant's warehouse facility is too narrow for easy access by some of the delivery vehicles. They can pass through, but the clearance is very tight. This is an annoying problem, but if the fix is quite costly, you could probably "live with it.") Questions you should ask early in the process are: What are the resources available to solve the problem? How many people can you allocate to the problem, and for how long a time? How soon do you need a solution? Today? Tomorrow? Next year? These are key questions to keep in mind as you take your first steps along the way to a problem solution. The quality of your solution is often, but not always, related to the time and money you have to *generate it and carry it through.* In some instances it may be necessary to extend deadlines in order to obtain a quality solution.

It may not be possible to completely address the cost issue until we are further along in the solution process. The cost will depend on whether or not the solution will be a permanent one or if it will be a temporary or patchwork solution. Sometimes *two* solutions are required: One to treat short-term symptoms to keep the process operating and one to solve the real problem for the long term. Be aware of these two mindsets in the problem-solving process. In some cases the **No's** in the figure on deciding if the problem should be solved can be changed to **Yes's** by *selling* the project to management. This change can be achieved by showing that the problem is an important one and is relevant to the operation of the company.

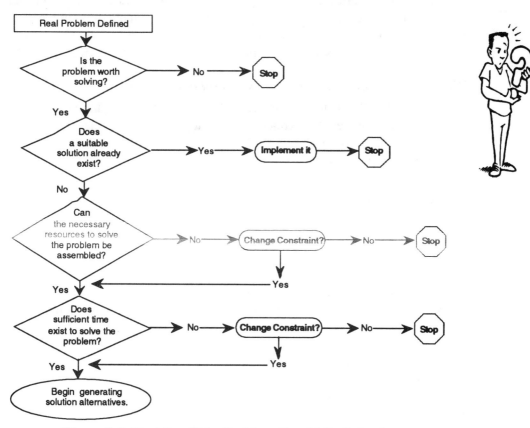

Figure 3-3. Deciding If the Problem Should Be Solved

Step 6. *Continue to gather information and search the literature.*

Gather as much information as possible by reading texts and literature related to the problem to learn the underlying fundamental principles and peripheral concepts. Literature searches are particularly helpful. Perhaps a closely related problem has already been solved. George Quarderer of Dow Chemical Company appropriately describes the idea of reinventing the wheel by his statement, *"Four to six weeks in the laboratory can save you an hour in the library."* The message is clear: Doing a bit of research into the background of the problem may save you hours of time and effort.

Search out colleagues who may have useful information and pertinent ideas. Have them play "**What if... ?**" with you; that is, "What if you did this?" or "What if I applied this concept?" Also have them play the devil's advocate and deliberately challenge your ideas. This technique stimulates creative interactions.

Step 7. *Form simple hypotheses and quickly test them.*

Returning to the Dead Fish example, an experienced problem solver could hypothesize that there was something else present in the water that was killing the fish. This hypothesis could be tested in the laboratory by analyzing samples of river water or by performing post-mortem examinations on the dead fish. These tests may have uncovered the presence of the fungus, thereby quickly defining the problem.

Step 8. *Brainstorm potential causes and solution alternatives.*

This last "first step" brings us to the close of the first phase of the creative problem-solving process and is really the first step of the second phase of the process: Generating Solutions to Problems. Techniques to generate solutions will be discussed in the next chapter.

Which Techniques to Choose

We do not expect the reader to apply every technique to every situation. In fact, when 400 problem solvers were surveyed as to which two techniques presented in this chapter were the most useful, *the choices were virtually equally divided among those presented in this chapter*. In other words, different techniques work better for different individuals and different situations, and it is a personal choice. The main point is to be organized as well as creative in your approach to problem definition.

SUMMARY

In this chapter we have discussed the necessity for defining the real problem. We have presented the eight steps that experienced problem solvers first use to attack problems. They are

- Collect and analyze information and data.
- Talk with people familiar with the problem.
- If at all possible, view the problem firsthand.
- Confirm all findings.
- Determine if the problem should be solved.
- Continue to gather information and search the literature.
- Form simple hypotheses and quickly test them.
- Brainstorm potential causes and solution alternatives.

Five problem definition techniques were presented to help you zero in on the true problem definition. They are

- *Find Out Where the Problem Came From*
  - Use the first four steps to gather information.
  - Learn who defined the problem initially.
  - Challenge reasoning and assumptions made to arrive at the problem statement given to you.

- *Explore the Problem*
  - Recall or learn the fundamental principles related to the problem.
  - Carry out an order-of-magnitude calculation.
  - Hypothesize what could be wrong.
  - Guess the result.

- *Present State/Desired State*
  - Write a statement of where you are and a statement of what you want to achieve and make sure they match.

- *Duncker Diagram*
  - Devise a pathway that makes it OK not to solve the problem posed to you.

- *Statement-Restatement*
  - Use the six triggers to restate the problem in a number of different ways.

REFERENCES

1. Kepner, C.H., and B. B. Tregoe, *The New Rational Manager*, Princeton Research Press, Princeton, NJ, 1981.
2. Woods, D.R., *A Strategy for Problem Solving*, 3rd ed., Department of Chemical Engineering, McMaster University, Hamilton, Ontario, 1985; *Chem. Eng. Educ.,* p. 132, Summer 1979; *AIChE Symposium Series,* 79 (228), 1983.
3. Higgins, J.S., et al., "Identifying and Solving Problems in Engineering Design," *Studies in Higher Education*, 14, No. 2, p. 169, 1989.
4. Parnes, S.J., *Creative Behavior Workbook*, Scribner, New York, 1967.

FURTHER READING

Copulsky, William, "Vision → Innovation," *Chemtech*, 19, p. 279, May 1989. Interesting anecdotes on problem definition and vision related to a number of popular products.

DeBono, Edward, "*Serious Creativity*," Harper Business, a division of Harper Collins Publishers, New York, 1993. Summary of 20 years of creativity researched by deBono. Many useful additional problem definition techniques are presented.

EXERCISES

1. Make a list of the most important things you learned from this chapter. Identify at least three techniques that you believe will change the ways you think about defining and solving problems. Which problem definition techniques do you find most useful? Prepare a matrix table listing all the problem definition techniques discussed in this chapter. Identify those attributes that some of the techniques have in common and also those attributes that are unique to a given technique.

	Attribute 1	Attribute 2	Attribute 3
Technique A	X		X
Technique B		X	X
Technique C	X	X	

2. Write a sentence describing a problem you have. Apply the triggers in the *Statement-Restatement Technique* to your problem.

    Perceived Problem Statement _____

    _____

    _____

    Restatement 1 _____

    _____

    _____

    _____

Restatement 2 _____

_____

_____

_____

Final Problem Statement _____

_____

_____

_____

Next apply the Duncker Diagram to this same problem. (Use the Duncker Diagram work sheet on page 59.)

3. Carry out a *Present State/Desired State* analysis on "I want a summer internship but no one is hiring" and then prepare a Duncker Diagram to solve the problem.

4. You have had a very hectic morning, so you leave work a little early to relax a bit before you meet your supervisor, who is flying into a nearby airport. You have not seen your supervisor from the home office for about a year now. He has written to you saying that he wants to meet with you personally to discuss the last project. Through no fault of yours, everything went wrong: The oil embargo delayed shipment of all the key parts, your project manager met with a skiing accident, and your secretary enclosed the key files in a parcel that was sent, by mistake, to Japan via sea mail. Your supervisor thinks that you have been so careless on this project that you would lock yourself out of your own car.

As you are driving through the pleasant countryside on this chilly late fall afternoon, you realize that you will be an hour early. You spot a rather secluded roadside park about 200 m away. A quiet stream bubbles through the park, containing trees in all their autumn colors. Such an ideal place to just get out and relax. You pull off into the park, absentmindedly get out and lock the car, and stroll by the stream. When you return, you find the keys are locked in the car. The road to the airport is not the usual route; there are cars about every 10 to 15 minutes. The airport is 9 km away; the nearest house (with a telephone) is 1 km away. The plane is due to arrive in 20 minutes. Your car, which is not a convertible, is such that you cannot get under the hood or into the trunk from the outside. All the windows are up and secured. Apply the Duncker Diagram and one other problem-solving technique to help decide what to do. (D.R. Woods, McMaster University)

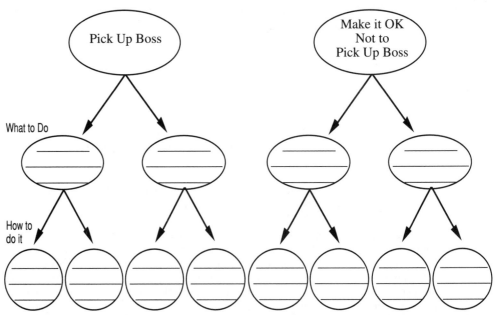

5. You are driving from Cambridge to London on the M11 motorway (expressway). You are scheduled to give a very important slide presentation at 1 PM. The drive normally takes 1 hr and 30 minutes but this morning you left at 10:30 AM to insure you had sufficient time. Suddenly your car stalls on the motorway halfway between Cambridge and London. What do you do? Apply two or more problem definition techniques to help answer this problem. (From J. Higgins and S. Richardson, Imperial College, London)

6. A propellant used in an air bag system is the chemical sodium azide. It is mixed with an oxidizing agent and pressed into pellets which are hermetically sealed into a steel or aluminum can. Upon impact, ignition of the pelletized sodium azide generates nitrogen gas that inflates the air bag. Unfortunately, if it contacts acids or heavy metal (e.g., lead, copper, mercury and their alloys), it forms toxic and sensitive explosives. Consequently, at the end of an automobile's life, a serious problem surfaces when an automobile with an undetonated airbag is sent to the junk yard for compacting and shredding, whereby it could contact heavy metals. The potential for an explosion during processing represents a serious danger for those operating the scrap recycling plant. Apply two or three problem definition techniques to this situation. (*Chemtech,* 23, p. 54, 1993)

7. Pillsbury, a leader in the manufacture of high-quality baking products, had its origins in the manufacture of flour for the baking industry. However, at the time Charles Pillsbury purchased his first mill in Minneapolis, the wheat from Minnesota was considered to be substandard when compared to the wheat used in the St. Louis mills, then the hub of the milling industry. Part of the problem was that winter wheat, commonly used in high-grade flour, could not be grown in Minnesota because of the long and cold winters. Consequently, the Minnesota mills were forced to use spring wheat which had a harder shell. At the time, the most commonly used milling machines used a "low grinding" process to separate the wheat from the chaff. The low grinding process refers to using stone wheels. A stone wheel rests directly on the bottom wheel, with the wheat to be

ground placed between them. With harder wheats, a large amount of heat was generated, discoloring and degrading the product quality. Thus, the flour produced from the Minnesota mills was discolored, inferior, and had less nutritional value and a shorter shelf life. The directions given could have been "Order more river barges to ship winter wheat up the Mississippi from St. Louis to Minneapolis." Apply two or more problem definition techniques to the situation. (Adapted from "When in Rome" by Jane Ammeson, *Northwestern Airlines World Traveler*, 25, No. 3, p. 20, 1993.)

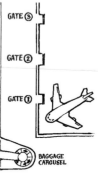

8. *Late Baggage.* An airline at the Houston Airport tried to please the passengers by always docking the plane at a gate within a one to two minute walk to the airport entrance and baggage claim and by having all the bags at baggage claim within eight to ten minutes. However, many complaints were received by the airline about the time it took to get the bags to the claim area. The airline researched the situation and found that there was virtually no way they could unload the bags to the transport trucks, drive to the unloading zone, and unload the bags any faster. However, the airline didn't change the baggage unloading procedure, but did change another component of the arrival process and the complaints disappeared. The airline did not use mirrors to solve the problem as was the case for the slow elevators. (a) What was the real problem? (b) Suggest a number of things that you think the airline might have done to eliminate the complaints. Apply two or more problem-solving techniques. (*The Washington Post,* p. A3, Dec. 14, 1992)

9. In 1991, 64% of all commercial radio stations in the country lost money. In order for a radio station to remain solvent it must have significant revenue from advertisers. Advertisers, in turn, target the market they consider desirable (i.e., income, spending, interest), and for the past several years this target has been the age group from 25 to 54. Along with the revenue loss, the number of radio stations playing the *Top 40* songs (i.e., the 40 most popular songs of that week) has decreased by a factor of 2 in the past three years, as did the audience for the *Top 40* songs. Many stations tried playing a blend of current hits with hits of 10 and 20 years ago; however, this blend irritated the younger listeners and also did not seem to solve the economic problem. Apply two or more problem definition techniques to this situation. (Adapted from *The International Herald Tribune*, p. 7, March 24, 1993.)

10. *The situation:* Sara is a freshman away at college preparing for her first final exams. She is homesick, stressed out, and would like to go home for the weekend to visit her parents, but her car is not working.

(**Present State**)   (**Desired State**)

Sara's car is not working.   At home with her parents.

Discussion: These states do not match and this mismatch confuses the problem. Which problem should she be attacking? The malfunctioning car? The visit?

First Revision.

(**Present State**)   (**Desired State**)

Cleaning up the Problem Statement

Continue in this manner until the states match.

11. *The situation:* FireKing is a small manufacturer of rich looking fireproof filing cabinets and wanted to increase its market share of 3%. While the designs were elegant, the cabinets were also the heaviest ones on the market and in people's minds, this meant the highest quality. However, higher weight meant higher shipping and transportation costs which made them very expensive. FireKing asked the following question, "How can we make our product lighter so as to have a competitive price?" However, some executives

believed a lighter product might hurt the image of quality. Apply one or more problem definition techniques to this situation. (David Turczyn)

12. ***The situation:*** A new method for killing roaches was developed by Bug-B-Gone Company which was more effective than any of the other leading products. In fact, no spraying was necessary because the active ingredient was in a container that is placed on the floor or in corners and the roach problem would disappear. This method has the advantage that product does all the work. The user does not need to search out and spray the live roaches. The product was test marketed to housewives in some southern states. Everyone who saw the effectiveness test results agreed the new product was superior in killing roaches. However despite a massive ad campaign, the standard roach sprays were still far outselling the new product. Apply one or more problem definition techniques to this situation. (David Turczyn)

13. ***The situation:*** A pneumatic conveyor is a device that transports powdered solids using air in the same manner that money is transported from your car at a bank's drive-through window. In the figure below, the solids are "sucked" out of the storage hopper and conveyed by air into the discharge hopper.
    The instructions given to solve the perceived problem: *"Find an easier way to clean a pneumatic conveying system when it plugs and interrupts operation."*

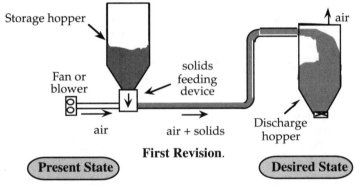

First Revision.

Present State

Conveying system plugs, interrupting operation.

Desired State

The system is easily and rapidly cleaned.

Continue in this manner until the states match.

14. ***The situation:*** A major American soap company carried out a massive advertising campaign to expand its market into Poland. The T.V. commercials featured a beautiful woman using the company's soap during her morning shower. Thousands of sample cakes were distributed door to door throughout the country. Despite these massive promotional efforts, the campaign was entirely unsuccessful. Polish television had been used primarily for communist party politics, and commericals were relatively rare. What is aired is usually party line politics. Apply one or more problem definition techniques to this situation. (Christina Nusbaum)

15. ***The situation:*** Employees are allowed to take merchandise out of the department store on approval. The original procedure required the employee to write an approval slip stating the merchandise taken. However, some employees were abusing the system by taking the clothing and destroying the slip, thereby leaving no record of the removed merchandise. Apply one or more problem definition techniques to this situation. (Maggie Michael)

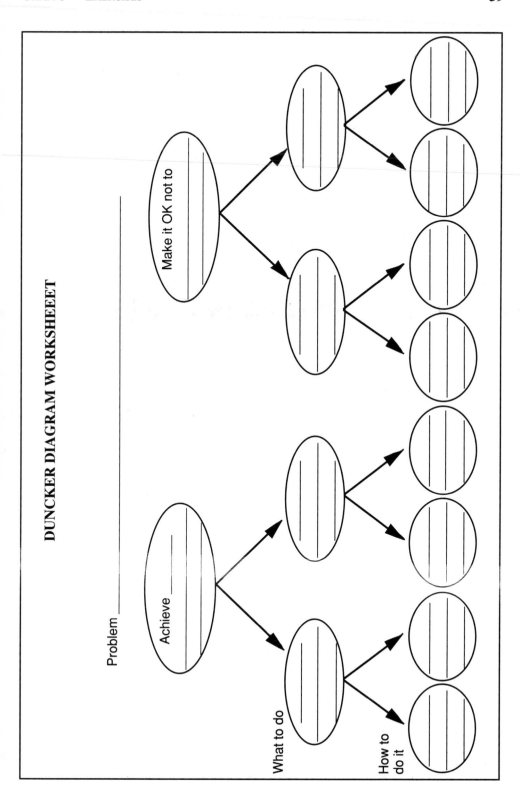

DUNCKER DIAGRAM WORKSHEET

Problem _____

Achieve _____

Make it OK not to

What to do

How to
do it

4 GENERATING SOLUTIONS

Nothing is more dangerous than an idea, when it is the only one you have.

- Emile Chartier

Once you have defined the problem you want to make sure you generate the best solution. Sometimes problems may seem unsolvable or they may appear to have only one solution, which as Emile Chartier points out is quite dangerous. This is a situation where you can use the idea generation techniques in this chapter to lead you to find the best solutions. Perseverance is perhaps the most notable characteristic of successful problem solvers, so you shouldn't become discouraged when solutions aren't immediately evident. Many times mental blocks hinder your progress toward a solution. The first step to overcoming these blocks is to recognize them, and then use blockbusting techniques to move forward toward the best solution.

What is the nature of these mental blocks and what causes them? Some common causes of blocks have been summarized by Higgins et al.:[1]

Define
Generate
Decide
Implement
Evaluate

Common Causes Of Mental Blocks

- Defining the problem too narrowly.
- Attacking the symptoms and not the real problem.
- Assuming there is only one right answer.
- Getting "hooked" on the first solution that comes to mind.
- Getting "hooked" on a solution that almost works (but really doesn't).
- Being distracted by irrelevant information, called "mental dazzle."
- Getting frustrated by lack of success.
- Being too anxious to finish.
- Defining the problem ambiguously.

There is a direct correlation between the time people spend "playing" with a problem and the diversity of the solutions generated. Don't be afraid to "play" with the problem. Let's look at how easy it is to have a conceptual block to a problem. Try this exercise *before* reading the several solutions provided on the following page.

Draw four or fewer straight lines (without lifting the pencil from the paper) that will cross through all nine dots. (Adams,[2] pp. 16–20)

● ● ●
● ● ●
● ● ●

The Nine Dot Problem

This puzzle is very difficult to solve if the imaginary boundary created by the eight outer dots is not crossed. Another common assumption that is not part of the problem statement is that the lines must go through the centers of the dots. Two possible solutions are provided below.

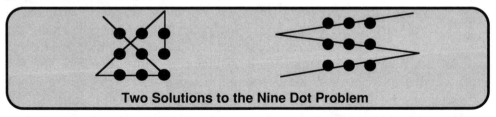

Two Solutions to the Nine Dot Problem

Several other creative solutions to the nine dot problem exist. These include rolling up the piece of paper such that it is cylindrical in shape and then drawing one line around the cylinder that passes through all nine dots, or photoreducing the nine dots and then using a thick felt pen to connect them with a single line. Another suggestion is to crumple up the piece of paper and stab it with a pencil (this is a statistical approach that may require more than one attempt to hit all the dots).

The purpose of this exercise is to show that putting too many constraints (either consciously or unconsciously) on the problem statement narrows the range of possible solutions. Normally, novice problem solvers will not cross a perceived imaginary limit—a constraint that is formed unconsciously in the mind of the problem solver—even though it is not part of the problem statement. Whenever you are faced with a problem, recall the nine dots to remind yourself to challenge the constraints.

4.1 RECOGNIZING MENTAL BLOCKS

Conceptual Blockbusting by James L. Adams[2] focuses on the cultivation of idea-generating and problem-solving abilities. The first step to becoming a better problem solver is to understand what conceptual blocks are and how they interfere with problem solving. A conceptual block is a mental wall that prevents the problem solver from correctly perceiving a problem or conceiving its solution. The most frequently occurring conceptual blocks are perceptual blocks, emotional blocks, cultural blocks, environmental blocks, intellectual blocks, and expressive blocks.

A. Perceptual Blocks are obstacles that prevent the problem solver from clearly perceiving either the problem itself or the information needed to solve it. A few types of perceptual blocks are
- *Stereotyping*

Survival training teaches individuals to make full use of all the resources at their disposal when they are faced with a life-threatening situation. For example, if you were stranded in the desert after the crash of your small airplane, you would have to make creative use of your available resources to survive and be rescued. Consider the flashlight that was in your tool kit. The *stereotypical* use for it would be for signaling, finding things in the dark, etc. But how about using the batteries to start a fire, the casing for a drinking vessel for water that you find in the desert cacti, or the reflector as a signaling mirror in the daylight, etc.

• *Limiting the problem unnecessarily*

The nine dot problem above is an example of limiting the problem *unnecessarily*. The boundaries of the problem must be explored and challenged.

• *Saturation or information overload*

Too much information can be nearly as big a problem as not enough information. You can become overloaded with minute details and be unable to sort out the critical aspects of the problem. Air traffic controllers have learned to overcome this block. They face information overload regularly in the course of their jobs, particularly during bad weather. They are skilled in sorting out the essential information to ensure safe landings and takeoffs for thousands of aircraft daily.

B. Emotional Blocks interfere with your ability to solve problems in many ways. They decrease the amount of freedom with which you explore and manipulate ideas, and they interfere with your ability to conceptualize fluently and flexibly. Emotional blocks also prevent you from communicating your ideas to others in a manner that will gain their approval. Some types of emotional blocks include:

• *Fear of risk taking*

This block usually stems from childhood. Most people grow up being rewarded for solving problems correctly and punished for solving problems incorrectly. Implementing a creative idea is like taking a risk. You take the risk of making a mistake, looking foolish, losing your job, or in a student's case, getting an unacceptable grade. In Chapter 2, some ideas for overcoming the fear of risk taking were discussed.

• *Lack of appetite for chaos*

Problem solvers must learn to live with confusion. For example, the criteria for the *best* solution may seem contradictory. What may be best for the individual may not be best for the organization or group.

• *Judging rather than generating ideas*

This block can stem from approaching the problem with a negative attitude. Judging ideas too quickly can discourage even the most creative problem solvers. Wild ideas can sometimes trigger feasible ideas which lead to innovative solutions. This block can be avoided by approaching the problem with a positive attitude.

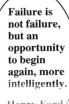

Failure is not failure, but an opportunity to begin again, more intelligently.

-Henry Ford

• *Lack of challenge*

Sometimes, problem solvers don't want to get started because they perceive the problem is too trivial and can be easily solved. They feel that the problem is not worthy of their efforts.

• *Inability to incubate*

Rushing to solve the problem just to get it off your mind can create blocks.

C. Cultural Blocks are acquired by exposure to a given set of cultural patterns, and environmental blocks are imposed by our immediate social and physical environment. One type of cultural block is the failure to consider an act that causes displeasure or disgust to certain members of society. To illustrate this type of block, Adams uses the following problem:[2]

Rescuing a Ping Pong Ball

Two pipes, which serve as pole mounts for a volleyball net, are embedded in the floor of a gymnasium. During a game of ping pong, the ball accidentally rolls into one of the pipes because the pipe cover was not replaced (see below). The inside pipe diameter is 0.06" larger than the diameter of a ping-pong ball (1. 50") which is resting gently at the bottom of the pipe. You are one of a group of six people in the gym, along with the following objects:

A 15' extension cord
A carpenter's hammer
A chisel
A bag of potato chips
A file
A wire coat hanger
A monkey wrench
A flash light

List as many ways as you can think of (in five minutes) to get the ball out of the pipe without leaving the gym, or damaging the ball, pipe, or floor.

Comment: A common solution to the problem is to smash the handle of the hammer with the monkey wrench and to use the splinters to obtain the ball. Another less obvious solution is to urinate in the pipe. Many people do not think of this solution because of a cultural block, since urination is considered a "private" activity in many countries.

> Being aware of potential conceptual blocks is the first step to overcoming them.

 Other types of mental blocks are

D. Environmental Blocks: Distractions (phones, easy intrusions) are blocks that inhibit deep prolonged concentration. Working in an atmosphere that is pleasant and supportive most often increases the productivity of the problem solver. On the other hand, working under conditions where there is a lack of emotional, physical, economical, or organizational support to bring ideas into action usually has a negative effect on the problem solver and decreases the level of productivity. Ideas for establishing a working environment that enhance creativity were presented in Chapter 2.

E. Intellectual Blocks: This block can occur as a result of inflexible or inadequate uses of problem-solving strategies. Lacking the necessary intellectual skills to solve a problem can certainly be a block as can lack of the information necessary to solve the problem. For example, attempting to solve complicated satellite communications problems without sufficient background in the area would soon result in blocked progress. Additional background, training, or resources may be necessary to solve a problem. Don't be afraid to ask for help.

 F. Expressive Blocks: The inability to communicate your ideas to others, in either verbal or written form, can also block your progress. Anyone who has played a game of charades or Pictionary™ can certainly relate to the difficulties that this type of block can cause. Make sketches, drawings, and don't be afraid to take time to explain your problem to others.

As we have just seen, there are many types and causes of mental blocks. If you find your problem-solving efforts afflicted by one of them, what can you do? Try one of the blockbusting techniques that we present next!

4.2 BLOCKBUSTING

A number of structured techniques are available for breaking through mental roadblocks.[3] Collectively, they are referred to as blockbusting techniques. Goman identifies a number of blocks to creativity and offers some suggestions on how to overcome these blocks.[4] The following table summarizes these blocks and blockbusters.

Goman's Blockbusters

Block	Blockbuster
1. NEGATIVE ATTITUDE Focusing attention on negative aspects of the problem and possible unsatisfactory outcomes hampers creativity.	1. ATTITUDE ADJUSTMENT List the positive aspects and outcomes of the problem. Realize that with every problem there is not only a danger of failure but an opportunity for success.
2. FEAR OF FAILURE One of the greatest inhibitors to creativity is the fear of failure and the inability to take a risk.	2. RISK TAKING Outline what the risk is, why it is important, what is the worst possible outcome, what your options are with the worst possible outcome, and how you would deal with this failure.
3. FOLLOWING THE RULES Some rules are necessary, such as stopping at a red light, while other rules hinder innovation.	3. BREAKING THE RULES Practice trying new things. Take a different route to work, try a new food, go somewhere you've never gone.
4. OVERRELIANCE ON LOGIC Relegate imagination to the background because of a need to proceed in a step-by-step fashion.	4. INTERNAL CREATIVE CLIMATE Turn the situation over to your imagination, your feelings, your sense of humor. Play with insights and possibilities.
5. YOU AREN'T CREATIVE Believing that you are not creative can be a serious hindrance to generating creative solutions. **Believing that you can't do something is a self-fulfilling prophesy** .	5. CREATIVE BELIEFS Encourage your creativity, by asking "what if" questions; daydream; make up metaphors and analogies. Try different ways of expressing your creativity.

In regard to Goman's fifth Blockbuster, Raudelsepp has presented definitive ways you can increase your creativity by learning new attitudes, values, and ways of approaching and solving problems by heeding the following principles.[5]

Improving Your Creative Abilities

• *Keep track of your ideas at all times*. Many times ideas come at unexpected times. If an idea is not written down within 24 hours it will usually be forgotten.

• *Pose new questions to yourself every day*. An inquiring mind is a creatively active one that enlarges its area of awareness.

• *Keep abreast of your field*. Read the magazines, trade journals, and other literature in your field to make sure you are not using yesterday's technology to solve today's problems.

• *Learn about things outside your specialty*. Use cross-fertilization to bring ideas and concepts from one field or specialty to another.

• *Avoid rigid, set patterns of doing things*. Overcome biases and preconceived notions by looking at the problem from a fresh view point, always developing at least two or more alternative solutions to your problem.

• *Be open and receptive to ideas (yours and others)*. Rarely does an innovative solution or idea arrive complete with all its parts ready to be implemented. New ideas are fragile; keep them from breaking by seizing on the tentative, half-formed concepts and possibilities and developing them.

• *Be alert in your observations*. This principle is a key to successfully applying the Kepner-Tregoe strategies discussed in the next chapter. Be alert by looking for similarities, differences, as well as unique and distinguishing features in situations and problems. The larger the number of relationships you can identify, the better your chances will be of generating original combinations and creative solutions.

• *Adopt a risk taking attitude*. Fear of failure is the major impediment to generating solutions which are risky (i.e., small chance of succeeding) but would have a major impact if they are successful. Outlining the ways you could fail and how you would deal with these failures will reduce this obstacle to creativity.

• *Keep your sense of humor*. You are more creative when you are relaxed. Humor aids in putting your problems (and yourself) in perspective. Many times it relieves tension and makes you more relaxed.

• *Engage in creative hobbies*. Hobbies can also help you relax. Working puzzles, playing games help keep your mind active. An active mind is necessary for creative growth.

• *Have courage and self-confidence*. Be a paradigm pioneer. Assume that you can and will indeed solve the problem. Persist and have the tenacity to overcome obstacles that block the solution pathway.

• *Learn to know and understand yourself*. Deepen your self-knowledge by learning your <u>real</u> strengths, skills, weaknesses, dislikes, biases, expectations, fears, and prejudices.

Dr. Edward de Bono, the international creativity authority, is serious about the need for creative thinking.[6] In his book *Serious Creativity*, deBono, the father of lateral thinking, takes the opportunity to summarize 25 years of research into creative thinking techniques.

Remember, one of the first steps in the problem solving process recommended by experienced problem solvers was the gathering of information. deBono cautions problem solvers in this regard. For example, when one begins working on a new problem or research topic, it is normal to read all the information available on the problem. To fail to do this may mean "reinventing the wheel" and wasting much time. However, during the course of information gathering, you may destroy your chances of obtaining an original and creative solution if you are not careful. As you read, you will be exposed to all the existing assumptions and prejudices that have been developed by previous workers or researchers. Try as you may to remain objective and original, your innocence will have been been lost. deBono recommends reading enough to familiarize yourself with the problem and get a "feel" for it. At this point you may wish to stop and organize some of your own ideas before proceeding with an exhaustive review of the literature. In this way you can best preserve your opportunities for creativity and innovation.

Have you ever heard the old saying: *"If it ain't broke, don't fix it."* deBono claims the attitude reflected by this statement was largely responsible for the decline of American industry. American managers operated in a strictly reactive mode, merely responding to problems as they arose. Meanwhile, the Japanese were fixing and improving things that weren't problems. Soon, the American "problem fixers" were left behind. To survive in today's business culture, proactive thinking, as opposed to reactive thinking, is required. This shift in thinking patterns requires creativity.

deBono summarizes a number of lateral thinking techniques that he popularized to improve creative thinking. These include random stimulation and the Six Thinking Hats. We will discuss random stimulation in this chapter. The Six Thinking Hats is an application of creative thinking that de Bono advocates for many situations. The hats are a device that help you look at a situation from many different viewpoints. As you imagine yourself wearing each different hat, you should assume the characteristics associated with that particular hat.

There are a variety of techniques that can be used to generate creative ideas. We will now begin to explore some of them.

4.3 BRAINSTORMING

> What one man is capable of conceiving, other men will be able to achieve.
> - Jules Verne

Brainstorming, one of the oldest techniques to stimulate creativity, is a familiar and effective technique for generating solutions. It provides an excellent means of getting the creative juices flowing. Recent surveys of people working in industry show that brainstorming is routinely used as an effective tool not only for one or two individuals discussing a problem in an informal setting but also in more formal large-group problem-solving sessions. Typically, the initial stages of idea generation begin with an unstructured free association of ideas to solve the problem (brainstorming). During this activity, lists of all possible solutions are generated either in group discussions or individually. The lists should include wild solutions or unusual solutions without regard to their feasibility. When brainstorming in groups, people can build upon each other's ideas or suggestions. This triggering of ideas in others is key to successful group brainstorming. Another critical component of group brainstorming is to maintain a positive group attitude. No negative comments or judgments are allowed during this stage of the solution process. Reserve evaluation and judgment until later. The more ideas that are generated, the better chance there is for an innovative, workable solution to the problem at hand. Nothing will kill a brainstorming session faster than negative comments. These comments must be kept in check by the group leader or the session will usually reduce to one of "braindrizzling."

Comments That Reduce Brainstorming to Braindrizzling

- That won't work.
- That's too radical.
- It's not our job.
- We don't have enough time.
- That's too much hassle.

- It's against our policy.
- We haven't done it that way before.
- That's too expensive.
- That's not practical.
- We can't solve this problem.

We conducted some brainstorming exercises with a number of groups of students. Some of the exercises were free-format in nature, totally unstructured, where the only guideline used was to generate as many ideas as possible. An example of an unstructured session is shown on the next page.

> **Problem Statement:** *How could the rules of basketball be changed so that players under 5'9" tall might be more competitive?*
>
> ### Ideas Generated:
>
> - Lower the height of the basket.
> - Two separate baskets.
> - Platform tennis shoes.
> - Tall players can't block.
> - Tall players can't rebound.
> - Tall players can't dribble.
> - Tall players can't jump.
> - No fouls on short players.
> - Tall players can't look at the basket.
> - Tall players can't use the backboard.
> - Play in zero gravity.
> - Some players on each team under 5'9".
> - Short players' baskets count more.
> - Taller players are not allowed outside the key.
> - 3-point shot line closer for shorter players.
> - Tall players can guard only tall players.
> - Tall players have to wear weighted shoes.
> - Short players can use trampolines.
> - Tall players must use a heavier ball.
> - Make tall players run (winded) before game.
> - Tall players wear uniforms with itching powder.
> - Allow players to pick other players up.
> - Short players wear spikes.
> - Tall players must carry a small child on their backs.
> - Tall players wear glasses restricting peripheral vision.
> - Short and tall teams: Short teams have more players.

Usually the ideas flow quickly at first and then slow abruptly after several minutes. The process has hit a "roadblock." These roadblocks hinder our progress toward a solution. Now let's use some other blockbusting techniques to help overcome some mental blocks and generate additional alternatives.

4.3A Osborn's Checklist

Osborn's Checklist techniques are used to generate additional alternatives that are related to those previously obtained. It is useful to help a group build on one another's ideas (i.e., piggyback). The checklist is shown in an abbreviated format in the following table.[7]

Osborn's Checklist for Adding New Ideas

Adapt? ....How can this (product, idea, plan, etc.) be used as is? What are other uses it could be adapted to?

Modify? ....Change the meaning, material, color, shape, odor, etc.?

Magnify? ......Add new ingredient? Make longer, stronger, thicker, higher, etc.?

Minify? ....Split up? Take something out? Make lighter, lower, shorter, etc.?

Substitute? ...Who else, where else, or what else? Other ingredient, material, or approach?

Rearrange? ..Interchange parts? Other patterns, layouts? Transpose cause and effect? Change positives to negatives? Reverse roles? Turn it backwards or upside down? Sort?

Combine? .....Combine parts, units, ideas? Blend? Compromise? Combine from different categories?

Mag**nify**

Minify

Rerraange

Com →←bine

Continuing with the basketball example . . .

Adapt? . . . . Smaller players can foul as many times as they want (rule adaptation). Assists by smaller players count as points.

Modify? . . . . Raise baskets for taller players (modify court). Tall players stay inside 3 point line.

Magnify? . . . . Short player's baskets worth 4 points (magnify score).

Minify? . . . . Tall player's shots worth 1 point (minify score).

Rearrange? . . . Separate leagues for taller and shorter players (rearrange grouping).

Stuck

Unstuck

4.3B Random Stimulation

Random Stimulation is a technique which is especially useful if we are stuck or in a rut.[8] It is a way of generating totally different ideas than previously considered and can "jump start" the idea generation process and get it out of whatever current rut it may be in.

The introduction of strange or "weird" ideas during brainstorming should not be shunned but instead should be encouraged. *Random Stimulation* makes use of a random piece of information (perhaps a word culled from the dictionary or a book (e.g., eighth word down on page 125), or a random finger placement on one of the words in the sample list below. This word is used to act as a trigger or switch to change the patterns of thought when a mental roadblock occurs. The random word can be used to generate other words that can stimulate the flow of ideas.

Examples of Random Stimulation Words

all, albatross, airplane, air, animals, bag, basketball, bean, bee, bear, bump, bed, car, cannon, cap, control, cape, custard pie, dawn, deer, defense, dig, dive, dump, dumpster, ear, eavesdrop, evolution, eve, fawn, fix, find, fungus, food, ghost, graph, gulp, gum, hot, halo, hope, hammer, humbug, head, high, ice, icon, ill, jealous, jump, jig, jive, jinx, key, knife, kitchen, lump, lie, loan, live, Latvia, man, mop, market, make, maim, mane, notice, needle, new, next, nice, open, Oscar, opera, office, pen, powder, pump, Plato, pigeons, pocket, quick, quack, quiet, rage, rash, run, rigid, radar, Scrooge, stop, stove, save, saloon, sandwich, ski, simple, safe, sauce, sand, sphere, tea, time, ticket, treadmill, up, uneven, upside-down, vice, victor, vindicate, volume, violin, voice, wreak, witch, wide, wedge, x-ray, yearn, year, yazzle, zone, zoo, zip, zap

---

For example, in the 5'9" basketball player brainstorming session, the word *humbug* was chosen at random from a book. *Humbug* brought to mind (i.e., led to (→)) the word *scrooge* which led to (→) *mean* which led to (→) *rough*, which resulted in the idea of *more relaxed foul rules for short players*. The goal of the pattern change allows the problem to be viewed from new perspectives not previously considered.

Example of Random Stimulation

Problem: Continuing the basketball example

Random word or concept:

Humbug - What ideas come to mind?

Humbug → Scrooge → mean → rough → more relaxed foul rules for short players.

Jealous → rage → short players may **taunt** tall players to distract them.

Industrial Example of Random Stimulation

Problem: Make toxic holding tank safe.

A large tank to hold toxic waste from a certain process is to be built. The problem is that the tank must be safe.

Choose a random word: Airplane

Airplane. - An airplane flies over the toxic waste tank. What if a plane were to crash into this tank, causing it to rupture? Now use this idea to find a feasible concern. If a plane may crash into the tank, what about a forklift or a waste delivery truck? This is a real concern that we must deal with. To make a long story short, it was decided to build a fence and dike around the tank to serve as a protection barrier. The benefit of random stimulation is that it allows the generation of a reasonable alternative that may not have been considered before.

4.3C Other People's Views

When approaching a problem that involves the thoughts and feelings of others, a useful thinking tool is *Other People's Views*, or OPV.[8] The inability to see the problem from various viewpoints can be quite limiting. Imagining yourself in the role of the other person allows you to see complications of the problem not considered previously. For example, consider an argument between the new store manager and an employee. The issue is the employee's desire to take two weeks vacation during the store's busiest period, the Christmas season. The manager's main concern is having enough help to handle the sales volume. The employee, however, has made reservations for an Antarctic cruise, one year in advance (with the former manager's approval), and stands to lose a lot of money if he has to cancel them. The problem does not have a solution yet, but by using OPV each person can

THE WAY I SEE IT...

see what the other person stands to gain or lose from the vacation, and each has a better understanding of the types of compromises the other person might be willing to make. Automotive engineers must be aware of many viewpoints to design a successful vehicle. They must consider the views of the consumers, the marketing personnel, management, the safety department, the financial people, and the service personnel. Failure to consider any of these groups' views could result in a failed product. Examples using this technique are shown below.

Example of Other People's Views

Problem: Continuing the basketball example

Owners: They like to win and fill the arena with fans. Game must be exciting. It must have some advantages for the coaches to want to have shorter players. Maybe consider a maximum cumulative height for the team, so that teams with very small players can have more very tall players, and a better chance of winning.

Fans: They like fast, exciting games with good ball handling, shooting, and slam dunks. Maybe we *do* need to lower the basket.

Short Players: Want big bucks and to play in the pros.

Tall Players: Don't want the game changed.

Another Example of Other People's Views

Problem: Space capsule burns upon entering the atmosphere.

Project Manager: Complete the project on time.

NASA Accountant: Solve problem but keep cost low.

Engineer: New material should not interfere with capsule performance.

Materials Scientist: Find a material that can handle the high temperature on re-entry.

Astronaut: Doesn't care about the capsule, wants to return alive.

Final solution: Allow the surface of the capsule to be destroyed, protecting the astronauts.

The Case of the Putrid Pond

Problem Statement: A very large (500,000 sq. ft. $\cong$ 10 football fields) sludge pond is part of a waste treatment plant. The liquid in the pond is very viscous and sticky. From time to time, unwanted floating objects (dead animals, branches, etc.) appear on the pond and must be removed. Unfortunately, covering the pond is not an option. Devise ways to solve the problem.

Brainstorming

Use a crane.

Large net over the pond.

Use a hovercraft.

Use a helicopter.

Build rail system above sludge pond.

Build a fence around the pond.

Osborn's Checklist

MODIFY

Change treatment process to eliminate sludge.

Add chemical to break down branches and dead animals.

Change properties of pond, so things sink, then dredge.

SUBSTITUTE

Build in desert/change location.

Substitute many tanks for pond.

Anaerobic digesters.

MAGNIFY-MINIFY

Shallow pond so people can wade.

Make narrow and deep, then cover.

REARRANGE

Grinder to cut everything up.

De-vegetate surrounding countryside.

Bring in vultures and scarecrows.

Other People's Views

ANIMALS	Food around the side of the pond. Electric fence. Add obnoxious odor to keep animals away.
PIGEONS	Scarecrow/predatory bird. Large fans around the side to blow birds in opposite direction.
PLANT MANAGER	Change the law → It is OK to have dead animals in the pond. Mechanical arm that grabs stuff from the pond.

Random Stimulation

WORD	
Latvia	Run process in different country; remote location for plant.
Custard Pie	Food → eat → algae → inject bacteria that digest floating debris.
Ski	Ski chair lift system across the pond to reach down and to pick off dead animals.

4.3D Futuring

Futuring is a blockbusting technique that focuses on generating solutions which currently may not be technically feasible but could be in the future. In futuring we ask questions such as: What are the characteristics of an ideal solution? What currently existing problem would make our jobs easier when solved, or would solve many subsequent problems, or would make a major difference in the way we do business? One of futurist Joel Barker's key ideas is that you should be bold enough to suggest alternatives that promise major advances, yet may only have a small probability of success.

The rules for futuring are relatively simple: Try to imagine the ideal solution without regard to whether or not it is technically feasible. Then begin by making statements such as... "If (this) _____ happened, it would completely change the way I do business." The University of Michigan's College of Engineering Commission on Undergraduate Education used futuring exercises to help formulate the goals and directions of engineering education for the 1990s and into the twenty-first century. The members of the commission were asked, "What do you see the student doing in 1999?" Some answers included: "I see the students using interactive computing to learn all their lessons. There are animations of processes where the students can change operating parameters and get instant visual feedback on their effect." "I see lecture halls where the lecturer is a hologram of the most authoritative and dynamic professor in the world on that particular topic." In futuring, you visualize the idealized situation that you would like to have and then work on devising ways to attain it.

How to Use Futuring

- Examine the problem carefully to make sure the real problem has been defined.

- Now, imagine yourself at some point in the future after the problem has been solved. What are the benefits of having a solution?

- "Look around" in the future. Try to imagine an ideal solution to the problem at hand without regard to technical feasibility. Remember, in the future, anything is possible.

- Make statements such as: "If only (this) _____ would happen, I could solve... ."

- Dare to change the rules! The best solutions to some problems are contrary to conventional wisdom.

Futuring - Sludge Pond Problem Revisited

- We don't have sludge in the future.
- Genetic engineering– dead plants and animals decay instantaneously.
- No waste products in the future.
- Change pond to gaseous or solid state.
- Use sludge as energy source.
- Use sludge as building material.
- Grow vegetation on pond.
- Heat source to boil.
- High frequency sound source that keeps animals away.
- Use sludge for roads.

Futuring– Useful Products from Cheese Waste

The waste products from cheese and yogurt plants are quite acidic and consequently cannot be discharged directly into lakes or rivers. This waste must be treated so that it can be safely discharged from the plant. One suggestion is to build a waste treatment facility that will neutralize the acid and kill the bacteria in the waste. It is important to keep the cost of the treatment materials, as well as the capital cost of the facility, at a minimum so that we do not severely impact the profits of the yogurt and cheese making. Let's try an exercise in futuring.

Let's imagine ourselves in the future, with a booming yogurt- and cheese-making business. Why is our plant doing so well? Our plant is very successful because there are no wasted materials in our operation. All our "waste" streams are being put to good use. What are we using them for? The main waste stream contains sugar and protein. What could we be using those for? Protein is an essential dietary requirement. We could be separating the protein and using it for human consumption (food additives) or animal feed supplements (more likely). What about the sugar? Could it be sold to someone as a raw material for another process? What kind of process? Sugars can be fermented, can't they? Perhaps we could be using the sugar to produce ethanol for a profit. What's left after removing the protein and the sugar? Could this material be landfill? But, landfilling is placing the material in the ground. Could we place it in the ground for a profit? What about placing it on the surface of the ground? Maybe it could be used as a fertilizer? Or perhaps as a biodegradable de-icing product for use on the roads? The de-icing idea is already being used in some cities.

Summary

- **Define the Real Problem**: The problem is not how to treat the waste but more generally what to do with it.

- **Imagine the Future**: Plant is profitable and has no adverse environmental impact.

- **Generate Solutions**: Success due to no waste production. All byproducts are recycled or sold.

4.4 ORGANIZING BRAINSTORMING IDEAS: THE FISHBONE DIAGRAM

Fishbone diagrams are a graphical way to organize and record brainstorming ideas. The diagrams look like a fish skeleton (hence their name). To construct a fishbone diagram the following procedure is used:

1. Write the real problem in a box (or circle) to the right of the diagram. Draw a horizontal line (the backbone) extending from the problem to the left side:

2. Brainstorm potential solutions to the problem.

3. Categorize the potential solutions into several major categories and list them along the bottom or top of the diagram. Extend diagonal lines from the major categories to the backbone. These lines form the basic skeleton of the fishbone diagram.

4. Place the potential solutions related to each of the major categories along the appropriate line (or bone) in the diagram.

A fishbone diagram for organizing the ideas for the putrid pond problem is shown below. The most difficult task in constructing a fishbone diagram is deciding the major categories to use for organizing the options. In this example, we have selected "Retrieval Equipment," "Process Changes/Redesign," and "Prevention." The ideas that were generated fall nicely into these categories. Other common categories used in fishbone diagrams are personnel, equipment, method, materials, and environment. This activity of sorting and organizing the information is a very valuable effort in the solution process.

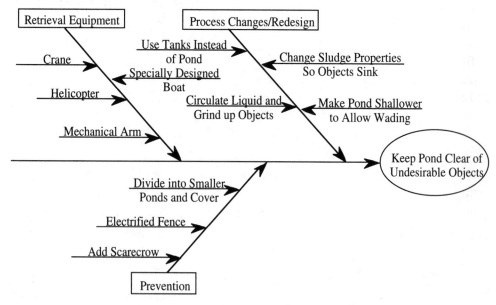

From the fishbone diagram above, we can evaluate the solutions that have been generated. We have put a structure to the solutions, organizing them and allowing us to "attack" the problem from a number of different fronts if we choose. These diagrams can be very helpful in visualizing all the ideas that you have generated.

4.5 BRAINWRITING

Two or more individuals are required in order to carry out an interactive brainstorming session. However, when there is no one to interact with, a technique being used by many companies is that of brainwriting. In brainwriting you follow the same procedure as brainstorming (e.g., free association, Osborn's checklist, random stimulation, futuring). Write down your ideas as fast as you generate them, never pausing or stopping to evaluate the idea. Also keep a notebook handy to write down ideas, because they often come at unusual times. After you have completed your list, organize your ideas (solutions) in a fishbone diagram.

4.6 ANALOGY AND CROSS-FERTILIZATION

It is well documented that a number of the most important advances in science, engineering, art, and business come from cross-fertilization and analogies with other disciplines. Here ideas, rules, laws, facts, and conventions from one discipline are transferred to another discipline. When we use analogies, we look for analogous situations/problems in other related and unrelated areas. Consequently, it is important that you read and learn about things outside your area of expertise. Generating ideas by *analogy* works quite well for many individuals. One recent example is that of Shockblockers™ Shoes developed by the U.S. Shoe Corporation (*Washington Post,* p. A47, December 18, 1992). The company wanted to develop shoes that absorb the shocks associated with walking. The company looked around to find out what other paraphernalia were used to protect the body from external contact. Ultimately they studied the materials inside a professional football helmet and eventually used the same shock absorbing foam in the soles in their new line of Shockblockers™ Shoes.

Remember the reentry of the space capsule problem in Chapter 3? One of the scientists used an analogy with a meteor entering the earth's atmosphere and asked why it did not burn up as a result of frictional heating. The answer was that the surface of the meteor was in a molten state and was being vaporized upon entering the earth's atmosphere. The frictional heat generated during reentry was dissipated into the heat of vaporization of the meteor surface. Consequently, the analogy between the space capsule and the meteor led to the use of a sacrificial material on the capsule surface that vaporized and thus dissipated the frictional heating.

In order to practice generating ideas by analogy and cross-fertilization, you might ask what each of the following pairs would learn from one another if they went to lunch or dinner together that would improve themselves, and/or the way they perform their job:

> A beautician and a college professor.
>
> A policeman and a software programmer.
>
> An automobile mechanic and an insurance salesman.
>
> A banker and a gardener.
>
> A choreographer and an air traffic controller.
>
> A maitre d' and a pastor.

Dinner at Antoine's

Let's consider a dinner meeting between a beautician and a college professor. The beautician could provide the professor with tips on the importance of having and maintaining a good physical appearance. Beauticians, also typically good conversationalists and listeners, could share these skills to help the professor establish a more effective rapport with the students. The professor would be better able to understand and respond to student concerns and problems. The professor might also pick up some tips on managing a small business which would be helpful in organizing and managing a research group.

College professors, on the other hand, are usually involved in research and are up on the latest developments in their field. The beautician could benefit from a discussion of these topics and be encouraged to obtain the newest beauty information and perhaps experiment with some new ideas. For example, new chemical/color treatments could be studied/explored using hair samples. The beautician could learn how to carry out an experiment by treating samples of hair with a new curling product for varying lengths of time to determine the optimum treatment procedure for different types of hair.

Many other combinations of professions would also provide growth experiences for both participants. The cross-fertilization of ideas from one group to another is a powerful method for adapting ideas from one discipline or profession to solve problems in another. Many times managers will bring together a small group of people from diverse backgrounds (ethnic, cultural) to interact and look at a problem and solution from many vantage points.

There are four steps you can use to solve problems by analogy:[4] 1) State the problem, 2) generate analogies (this problem is like trying to. . .), 3) solve the analogy, and 4) transfer the solution to the problem. When generating analogies, apply the same rules you did in brainstorming. For example in the case of the stale cereal, one could say, "Keeping the cereal fresh is like preserving raw fish in the tropics without a refrigerator and without cooking." How could one preserve fish? Add lemon/lime juice to make seviche (pickled fish). What could be added to the cereal to keep it fresh?

A Cold Winter's Day

The Situation: A large office building in the city was not as energy efficient as the building's owners would have liked. As a result, in order to keep their heating bills down, the building was kept colder than the occupants preferred, and many complaints were received.

Step One: State Problem *(What is the situation?)*
Occupants of building are too cold. Utilities bills are too high. Too many complaints.

Step Two: Generate Analogies *(What else is **like** this situation?)*
Generate as many possibilities as you can, then choose one to work with: Being cold in the office is like. . .

Being too cold at a football game.
Being too cold on a camping trip.
Being too cold in a car that hasn't warmed up in the winter.
Being too cold in bed at night.

Step Three: Solve the Analogy
When you are too cold on a camping trip, you build a campfire which serves as a source of both heat and light.

Step Four: Transfer the Solution to the Problem
Instead of building a campfire in the office, rent or buy a portable space heater. Use a readily available source of heat and light to solve the building's energy problems. Install a heat recovery system to recover waste heat from the fluorescent lights to warm the offices and improve the energy efficiency. (This is a practice that is used in modern energy-efficient office buildings.)

4.7 INCUBATING IDEAS

The incubation period is very important in problem solving. Working on a solution to a problem to meet a deadline often causes you to pick the first solution that comes to mind and then "run with it," instead of stopping to think of alternative solutions.

Once the generation of ideas has halted (or you collapse from the effort), an incubation period may be in order. Little is truly understood about mental incubation, but the basic process involves stopping active work on the problem and letting your subconscious continue the work. Everyone has, at one time or another, been told to "sleep on a problem," and maybe the solution will be apparent in the morning. This incubation or subconscious work has been described as a mental scanning of the billions of neurons in the brain in search of a novel or innovative connection to lead to a possible solution.[9] A number of members of the National Academy of Engineering were asked, "What do you do when you get stuck on a problem?" Some of the responses were

- "Communicate with other people. Read articles. Try new techniques *after a period of digestion*. Follow a lead if it looks promising. Keep pursuing."

- "Ask questions about all the circumstances. *Go home and think*. Go to your arsenal of past experiences. Identify factors related to the problem. Read, write and exchange ideas."

- "I write down everything that I must know to have a solution and everything that I know about the problem so far. Then I usually *let it sit overnight,* and think about it from time to time. While it is sitting I often review the recent literature on similar problems and often get an idea on how to proceed."

- "When I can afford the liberty of doing so, I will *put the problem down and do something else for awhile*. My mind keeps working on the problem, and often I will think of something while trying not to."

The common thread that runs through these responses is the notion of an incubation period. If the solution to the problem is not an emergency, incubation is a useful (in)activity to consider.

CLOSURE

The goal of this chapter was to present techniques to help you generate creative solutions. Mental blocks and techniques to remove them (blockbusting techniques) were presented. Blockbusting techniques help break preconceived notions about the problem situation. Many times it is advantageous to take a break when working on a problem to let your ideas incubate while your subconscious works on it. However, don't turn the responsibility over to your subconscious completely by saying, "Well, my subconscious hasn't solved the problem yet."

SUMMARY

- Be able to recognize the different mental blocks when they appear (Perceptual, Emotional, Cultural, Environmental, Intellectual, and Expressive Blocks).

- Use Goman's Blockbusters:
  Attitude Adjustment, Risk Taking, Breaking the Rules, Internal Creative Climate, and Creative Beliefs.

- Use Osborn's Checklist to generate new ideas:
  Adapt, Modify, Magnify, Minify, Rearrange, Combine.

- Use *Random Stimulation* and *Other People's Views* to generate new ideas when you are stuck in a rut.
  Telegraph → wire → electricity → light bulb → new ideas

- Remove all technical blocks to envision a solution in the future.

- Use a fishbone diagram to help organize the ideas/solutions you generate.

- Use analogy and cross-fertilization to bring ideas, phenomena, and knowledge from other disciplines to bear on your problem.

- Let the problem incubate so that your mind keeps working on it while you are doing other things.

REFERENCES

1. Higgins, J.S., et al., "Identifying and Solving Problems in Engineering Design," *Studies in Higher Education*, 14, No. 2, p. 169, 1989.

2. Adams, James L., *Conceptual Blockbusting: A Guide to Better Ideas*, W. H. Freeman and Company, San Francisco, 1974.

3. Van Gundy, A.B., *Techniques of Structured Problem Solving*, 2nd ed., Van Nostrand Reinhold, New York, 1988.

4. Goman, Carol K., *Creativity in Business–A Practical Guide for Creative Thinking*, Crisp Publications, Inc., 1200 Hamilton Ct., Menlo Park, CA 94025, 800-442-7477, 1989.

5. Raudelsepp, E., *Chemical Engineering*, 85, p. 95, July 2, 1979.

6. deBono, Edward, *Serious Creativity*, Harper Business, a division of Harper Collins Publishers, New York, 1993.

7. Felder, R.M., "Creativity in Engineering Education," *Chemical Engineering Education*, 22(3), 1988.

8. deBono, E., *Lateral Thinking*, Harper & Row, New York, 1970.

9. Reid, R. C., "Creativity?," *Chemtech*, 17, p. 14, January 1987.

FURTHER READING

Adams, James L., *Conceptual Blockbusting, A Guide to Better Ideas*, 3rd ed., Addison-Wesley
    Publishing Co., Inc., Stanford, CA, 1986.
von Oech, Roger, *A Whack on the Side of the Head, How You Can be More Creative*, revised
    edition, Warner Books, New York, 1990.

EXERCISES

1. Keep a journal of all the good ideas you generate.

    A. _____

    B. _____

    C. _____

2. a) Make a list of the worst business ideas you can think of (e.g., a maternity shop in a
    retirement village, a solar-powered night-light, *reversible* diapers).

    A. _____

    B. _____

    C. _____

    b) Take the list you generated in part (a) and turn it around to make them viable concepts
    for entrepreneurial ventures, (e.g., reversible diapers–blue on one side and pink on
    the other).

    A. _____

    B. _____

    C. _____

3. Apply Goman's four steps of generating solutions to problems by analogy to a problem
    you have.

    1. State the Problem.

    _____

    _____

    _____

    2. Create Analogies: This situation is like... .

        1. _____

        2. _____

        3. _____

    3. Solve the Analogy.

        1. _____

        2. _____

        3. _____

4. Transfer the Solution.

    1. _____

    2. _____

    3. _____

4. Rent a video. Watch half the movie with a friend(s). Stop the movie and each of you "create" your own ending. Watch the rest of the movie and discuss the results. Whose ending was better? Why?

5. Suggest 50 ways to increase spectator participation at a) professional basketball games. Examples: Have a drawing at each game and the people in the randomly selected seats get to play for two minutes. Give the fans one arrow each to shoot at the basketball in midair to try to block the shot. Suggest 50 ways for spectator participation in professional b) football, c) baseball, d) hockey.

6. Suggest or devise 50 different ways to cross a lake of molasses.

7. You are a passenger in a car without a speedometer. Describe 20 ways to determine the speed of the car.

8. An epidemic on a chicken farm created a thousand tons of dead chickens. The local landfill would not accept the dead chickens. It is also against the law to bury the chickens and the local authorities are insisting the matter be dealt with immediately. Suggest ways to solve the farmer's problems. (*Chemtech*, 22, 3, p. 192, 1992)

9. A reforestation effort in Canada is running into trouble in a particular region. In one nursery alone, 10 million seedlings were eaten by voles. The voles even consumed the varieties of seedlings chosen for the unpalatable phenol/condensed tannin secondary metabolite they contain. The voles overcame this unpalatability by cutting the branches, stripping the bark, and then leaving them for a few days before eating. This process caused the unpleasant components to decline to acceptable levels. Suggest 15 ways to solve the reforestation problem in this nursery. (*Chemtech*, 21, p. 324, 1991)

10. Kite flying is a growing hobby around the world. (They are very entertaining; it is not unusual to find kites that fly at altitudes of more than 2,000 feet.) Suggest 50 ways kites can be used for purposes other than entertainment.

11. a) Rearrange four pencils to make six equal triangles.

    b) Remove six pencils to leave two perfect squares and no odd pencils.

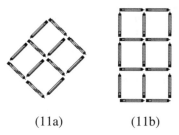

(11a)   (11b)

12. a) Rearrange three balls so that the triangle points up instead of down.

   b) Moving one black poker chip only, make two rows of four.

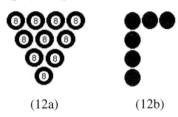

(12a)   (12b)

13. Apply a variety of brainstorming techniques to one or more of the following situations.

   a) Suggest ways to measure the pressure over the top of the carbonated liquid in a 2-liter bottle. This might be required to determine the necessary pressure rating for a bottle. How could you measure the pressure inside a balloon?

   b) A coin collector has a coin that she suspects is zinc. Suggest ways to determine nondestructively and precisely whether the coin is zinc. (Be specific.)

   c) Imagine yourself in the year 2020. What would an automobile look like? What would be some of the selling features?

   d) Suggest some ways to prevent the problem of driving under the influence of alcohol in the future.

   e) What features would be nice to have on a television, ten years from now? How about on a computer?

   Prepare a fishbone diagram of any one of the above brainstorming examples.

14. Choose two people from different professions (e.g., repairman, florist, dentist, accountant, policeman, hockey coach, car designer, custodian, bell hop, cruise ship activity director, cub scout leader) and make a table similar to the one below of what they could learn from one another that would enrich each others lives. (Matt Latham, Sue Stagg)

Pastor gives to a Maitre d'

1) ideas to rapidly assess people's needs.
2) suggestions on how not to take every problem he hears personally (thick skinned).
3) importance of good physical appearance.
4) suggestions on how far you can push people (in terms of views and ideals).
5) ideas on offering suggestions and advice.
6) ideas on how to be more self-reliant (scheduling).

Maitre d' gives to a Pastor

1) knowledge to calm upset individuals/ crowd control.
2) understanding and dealing with people, approachability.
3) memory techniques to remember frequent customers.
4) an appreciation of having a boss and someone watching what you do.
5) ideas on how to learn to be happy with your job and yourself.

15. Make a list of several ways you can improve your creative abilities. Describe how you would implement some techniques from the table on page 66.

A. _____   C. _____
   _____   _____
   _____   _____
B. _____   D. _____
   _____   _____
   _____   _____

16. Fifty-seven sticks are laid out to form the equation. Remove eight sticks to make the answer correct. Do not disturb any sticks other than the eight to be removed. First list any perceived constraints that you initially thought could be blocks to solving this problem. (Source: *Brain Busters* by Phillip J. Carter and Ken A. Russel, Sterling Publishing, Inc., New York, 1992)

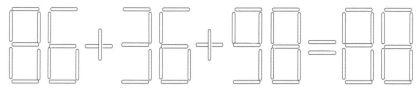

17. Carry out a futuring exercise to visualize
    a) A telephone call in the year 2010.
    b) Eating a meal with your family in the year 2050.
    c) A homework assignment in the year 2025.
    d) A homework assignment in the year 2125.

5 DECIDING THE COURSE OF ACTION

Once the real problem(s) is defined and we have generated a number of possible solutions, it is time to make some decisions. Specifically, we must

- Decide which problem to work on first
- Choose the best alternative solution
- Decide how to successfully implement the solution

An organized process for making these essential decisions is the Kepner-Tregoe (K.T.) Approach, which is described in *The New Rational Manager*.[1,2]

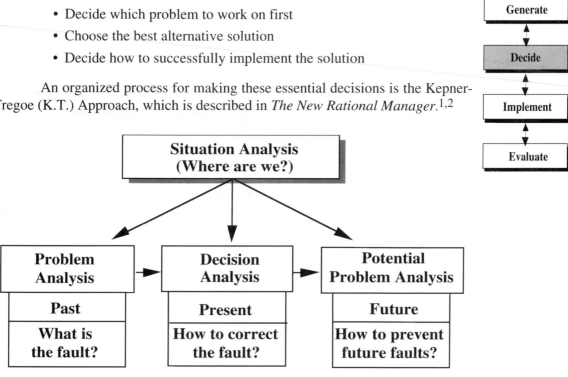

Figure 5-1. Components of the Kepner-Tregoe Approach

K.T. *Situation Analysis* not only helps us decide which problem to work on first; it also guides us with respect to what is to be done. Do we need to learn the cause (Problem Analysis, PA), make a decision (Decision Analysis, DA), or plan for success (Potential Problem Analysis, PPA)? That is, in situation analysis we also classify the problem into one of these analysis groups. In *Problem Analysis* the cause of the problem or the fault is unknown and we have to find it. What is it that happened in the *past* that is causing the current trouble? While the K.T. Problem Analysis might have fit more appropriately into Chapter 3 on Problem Definition, we believe it is best to present the entire K.T. Approach intact. In *Decision Analysis* the cause of the problem has been found and now we need to decide what to do about it. The decision at the *present* time is how to correct the fault. In *Potential Problem Analysis* we want to ensure the success of the decision and anticipate and prevent *future* problems from occurring.

Past Present

5.1 SITUATION ANALYSIS

Problem
solvers
must juggle
priorities
all the time.

In many situations, a number of problems arise at the same time. In some cases they are interconnected; in other cases, they are totally unrelated, and it is "just one of those days." When these situations occur, Kepner-Tregoe (K.T.) Situation Analysis can prove useful in helping to decide which problem receives the highest priority.

We first make a list of all of the problems and then try to decide which problem in this group should receive attention first. Each problem will be measured against three criteria: (1) timing, (2) trend, and (3) impact, each of which will be evaluated as being of a high (H), moderate (M), or low (L) degree of concern. We also decide what type K.T. analysis is to be carried out: PA, DA, or PPA.

5.1A Evaluation Criteria

1. *Timing*: How urgent is the problem? Is a deadline involved? What will happen if nothing is done for a while? For example, if one of the five ovens in a bakery is malfunctioning and the other four ovens could pick up the extra load, it may be possible to wait on this problem and address more urgent problems, so we would give the problem an L rating (low degree of concern). On the other hand, if the other four ovens are operating at maximum capacity and a major order must be filled by the evening, the rating for *timing* would be H (high degree of concern) because the problem must be solved now.

2. *Trend:* What is the problem's potential for growth? In the bakery example, suppose the malfunctioning oven is overheating, getting hotter and hotter, and cannot be turned off. Consequently the *trend* is getting worse, and you have a high degree of concern (an H) about a fire starting. You also could have a high degree of concern if you are getting further and further behind on your customer's orders. On the other hand, if the oven is off and you can keep up with the orders with four ovens, the *trend* is a low degree of concern (L).

3. *Impact:* How serious is the problem? What are the effects on the people, the product, the organization, and its policies? In the bakery example, suppose you cannot get the oven repaired in time to fill the order of a major client. If, as a result, you could subsequently lose the client's business, then the *impact* is a high degree of concern (H). On the other hand, if you can find a way to fill all the orders for the next few days, then the *impact* of one malfunctioning oven is a moderate degree of concern (M).

We now consider several examples and solutions to help illustrate the K.T. approach to prioritizing problems. First let's consider the problem of the man pictured on the next page.

K.T. Situation Analysis of :
You know it's a really bad day when...

Project	Timing	Trend	Impact	Process
1. Get dog off leg	H	H	H	DA
2. Repair car	L	L	M	PA
3. Put out fire	H	H	H	DA
4. Ensure papers in briefcase will not be destroyed	M	M	H	PPA
5. Prepare for touchdown of tornado	M	H	H	DA/PPA

1. It is necessary to get the dog off your leg now (High Priority). The trend is getting worse because there are more and more lacerations (High Priority) and the impact is that you can do nothing else until the dog is off your leg (High Priority). The process is to decide how to get the dog off your leg (DA).

2. Repairing the car can wait (Low Priority) and it is not getting worse (Low Priority), but if it is not repaired soon it could have impact on your job by your not being able to visit clients (Moderate Priority). The problem is to find out what is wrong with the car (PA).

3. Putting out the fire receives high priority in all three categories. The problem is to decide (DA) how to do it: Get the hose or fire extinguisher; call the fire department; and/or make sure everyone is out of the house.

4. If you rush off to handle the other projects in this list, you need to make sure your months of work, which includes signed documents in your briefcase, are protected. The process is one of Potential Problem Analysis (PPA) and of making sure your signed papers (which your clients now wish they had not signed) are in a safe place.

5. While the tornado looks somewhat close in the picture, it may be used to represent a tornado in the area, and thus may only be a tornado warning. So this hazard could merit Decision Analysis/Potential Problem Analysis.

First Day on the Job. . . Trial by Fire

Sara Brown just became manager of Brennan's Office Supply Store. The Brennan Company owns ten such stores in the Midwest. Sara's store, which is located in the downtown area on a busy street, has an inventory of over one million dollars and over 20,000 square feet of floor space. On her first day of work, Sara is inundated with problems. A very expensive custom-ordered desk that was delivered last week received a number of scratches during unpacking, and the stockroom manager wants to know what he should do. She just discovered that the store has not yet paid the utility bills that were due at the end of last month, and she realizes that the store has been habitually late paying its bills. The accounts receivable department tells her that it has had an abnormally high number of delinquent accounts over the past few months, and it wants to know what action should be taken. There is a large pile of boxes in the storeroom from last week that have yet to be opened and inventoried. The impression she has been getting all morning from the 30 employees is that they are all unhappy and dislike working at the store. To top things off, shortly after lunch, a large delivery truck pulls up to the front of the store and double-parks, blocking traffic. The driver comes into the store and announces that he has a shipment of 20 new executive desks. Where does Sara want them placed? The employees tell Sara that this shipment was not due until next week and there isn't any place to put them right now. Outside she hears horns of the angry drivers as the traffic jam grows. What should Sara do?

Situation Analysis

Major concern	Subconcern	Timing	Trend	Impact	Process
Space	Unopened Boxes	L	L	L	DA
	20 New Desks	H	H	H	DA
Personnel	Employee Morale	M	M	H	PA
Finances	Money Owed	M	M	H	DA
	Money Due	M	M	M	PA
Quality	Scratched Desk	L	L	M	DA/PPA

While boxes on the floor may be an eyesore and awkward to step around, it is not necessary we do anything about them immediately (L in timing). The situation is not getting worse by having them there (L in trend), and the impact of not having them opened and the contents shelved is low. The process to address this subconcern is decision analysis– we have to decide who is to open the boxes and when to do it. What to do about the 20 new desks has to be decided (DA) immediately and thus is a high degree of concern. The impact of not accepting or accepting and storing such a large order is a high degree of concern. A traffic jam is beginning to form and is getting worse while Sara is deciding what to do so the trend is a high (H) degree of concern. The employee morale needs to be addressed in the very near future. It is believed that lack of care and sloppiness were factors in damaging the custom-ordered desk, so its impact has a high degree of concern. The morale, while low, could get worse and therefore the trend is a moderate (M) degree of concern. We don't know why the moral is low so we need to carry out a problem analysis (PA)

- continued -

> First Day on the Job... Trial by Fire –continued
>
> to learn the problem. Sara needs to pay the utility bills fairly soon or the electrical power to the store could be shut off, which would cause a high degree of concern in the impact category. Sara needs to find out why the money due her has not been paid (Problem Analysis). Nothing needs to be done with the scratched desk immediately, but we do need to decide what to do in the not too distant future (DA). We also need to plan how to unpack the desks and other items more carefully (Potential Problem Analysis).

5.1B The Pareto Analysis and Diagram

When it is evident that there is more than one problem to be dealt with, a Pareto Analysis is another helpful tool for deciding which problems to attack first. This tool is commonly used in industry for quickly deciding which problem to attack first. The Pareto Analysis shows the *relative* importance of each individual problem to the other problems in the situation. Pareto Analysis draws its name from the Pareto Principle which states that 80% of the trouble comes from 20% of the problems. Thus, it helps to highlight the *vital few* concerns as opposed to the *trivial many*. The defects to investigate first for corrective action are those that will make the largest impact. As an example, let's consider the problems that the Toasty O's plant had with their product last year (See To Market, To Market example in Chapter 3). The problems were classified as follows:

		Number of Boxes
A.	Inferior printing on boxes (smeared/blurred)	10,000
B.	Overfilling boxes (too much weight)	30,000
C.	Boxes damaged during shipping	2,000
D.	Inner wrapper not sealed (stale)	25,000
E.	No prize in box	50,000

The data are shown graphically below:

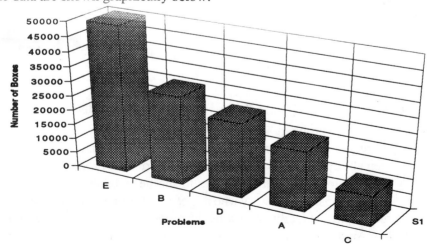

When the bar graph has the frequencies arranged in a descending order, the resulting figure is called a Pareto Diagram. Based on the number of boxes affected, the Toasty O's plant would probably attack the problem in the following order E-B-D-A-C. But, if they reexamine the data in terms of lost revenue instead of the number of boxes affected, a different picture of the problems emerges.

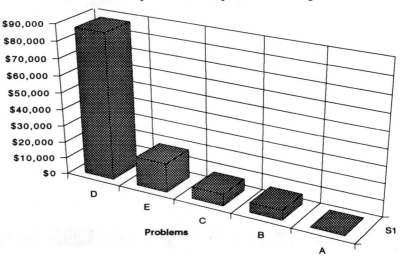

Boxes		Lost Revenue($)
A. Inferior printing on boxes (smeared/blurred)	10,000	$100
B. Overfilling boxes (too much weight)	30,000	$6,000
C. Boxes damaged during shipping	2,000	$7,000
D. Inner wrapper not sealed (stale)	25,000	$87,500
E. No prize in box	50,000	$17,500

From this graph it is clear that we can make the biggest impact on the problem situation by attacking the stale cereal problem (D) first, followed by E-C-B-A. When a Pareto Diagram is made, care should be taken to "weight" the problems using the most relevant quantity to the particular situation. In this case (and in many others) the impact on plant revenue is the key parameter. Pareto Diagrams are merely a useful, convenient way to organize and visualize problem data to help decide which of multiple problems to attack first.

5.2 K.T. PROBLEM ANALYSIS AND TROUBLESHOOTING

Our studies on problem-solving techniques in industry revealed that one of the major differences between experienced problem solvers and novice problem solvers was their ability to ask the right questions. Experienced problem solvers have learned to ask questions that will penetrate to the heart of the problem and to interview as many people as necessary who might have useful information about the problem. A technique that facilitates asking the proper questions is Kepner-Tregoe (K.T.) Problem Analysis. In this technique, *distinctions* are made between

- What *is* the problem and what *is not* the problem?
- Where did the problem occur? Where is everything OK?
- When did the problem first occur? When was everything OK?
- What is the magnitude (extent) of the problem?

This analysis is most useful in *troubleshooting operations* where the cause of the problem or fault is not known. Problems that lend themselves to K.T. Problem Analysis are ones in which an undesirable level of performance can be observed and compared with the accepted standard performance. For example, consider the following case in which a company ordered and received a new shipment of company stationery with the logo printed at the top. A few days later, it was noticed the logo was easily smeared. This smearing had never been observed before. In the K.T. analysis, Table 5-1, the deviation is that the printing quality was unacceptable and hence a problem must be precisely identified, described, and located.

TABLE 5-1: The Four K.T. Dimensions of a Problem

		IS	IS NOT	DISTINCTION	CAUSE
What:	Identify:	What is the problem?	What is not the problem?	What is the distinction between the **is** and the **is not**?	What is a possible cause?
Where:	Locate:	Where is the problem found?	Where is the problem not found?	What is distinctive about the difference in locations?	What is the possible cause?
When:	Timing:	When does the problem occur?	When does the problem not occur?	What is distinctive about the difference in the timing?	What is a possible cause?
		When was it first observed?	When was it last observed?	What is the distinction between these observations?	What is a possible cause?
Extent:	Magnitude:	How far does the problem extend?	How localized is the problem?	What is the distinction?	What is a possible cause?
		How many units are affected?	How many units are not affected?	What is the distinction?	What is a possible cause?
		How much of any one unit is affected?	How much of any one unit is not affected?	What is the distinction?	What is a possible cause?

A good problem
statement often
includes:
(a) What is known.
(b) What is unknown.
(c) What is sought.

The basic premise of K.T. Problem Analysis is that there is always something that distinguishes what the problem *IS* from what it *IS NOT*. The cause of the problem is usually a change that has taken place to produce undesirable effects. Things were OK, now they're not. Something has changed. (The printing company changed to a glossier paper.) The possible causes of the problem (deviation) are deduced by examining the differences found in the problem. (It is difficult to impregnate glossy paper with ink using the current printing process.) The most probable cause of the problem is the one which best explains all the observations and facts in the problem statement. (The ink is not penetrating the paper and thus it wipes off when used.)

Is or is not?
That is the question,
Watson!!

The real challenge is to identify the distinction between the *IS* and the *IS NOT*. Particular care should be taken when filling in the distinction column. Sometimes the distinction statement should be rewritten more than once in order to sharpen the statement to specify the distinction exactly. For example, in one problem analyzed by the K.T. method, the statement "two of the filaments were clear (OK) and two were black (not OK)" was sharpened to "two filaments were clear and two were covered with carbon soot." This *sharpening* of the distinction was instrumental in determining the reason for the black filament. Think in terms of dissimilarities. What distinguishes *this* fact (or category) from *that* fact (or category)? By examining the distinctions, possible causes are generated. This step is the most critical in the process and usually requires careful analysis, insight, and practice to ferret out the differences between the *IS* and *IS NOT*. From the possible causes, we try to ascertain the most probable cause. The most probable cause is the one that explains each dimension in the problem specification. The final step is to verify that the most probable cause is the true cause. This may be accomplished by making the appropriate change to see if the problem disappears.

In addition to what, when, where, and to what extent, it can sometimes be beneficial to add who, why, and how. For example,

> Who was involved?
>
> Who was not involved?
>
> Why is it important?
>
> Why is it not important?
>
> How did you arrive at this conclusion?

Troubleshooting is an important skill for problem solvers. Some guidelines for troubleshooting have been given by Woods.[3] The problem solver should also separate people's observations from their interpretations of what went wrong. A common mistake is to assume that the most obvious conclusion or the most common is always the correct one. (This is, however, a good place to start, though not necessarily to stop.) A famous medical school proverb that relates to the diagnosis of disease is: "When you hear hoofbeats, don't think zebras." In other words, look for common explanations first. Finally, the problem solver should continually reexamine the assumptions and discard them when necessary.

> When you hear hoofbeats, don't think zebras!

Fear of Flying.....

A new model of airplane was delivered to Eastern Airlines in 1980. Immediately after the planes were in operation, the flight attendants developed a red rash on their arms, hands, and faces. It did not appear on any other part of the body and the rash occurred only on flights that were over water. Fortunately, it usually disappeared in 24 hours and caused no additional problems beyond that time. When the attendants flew other planes over the same routes, no ill effects occurred. The rash did not occur on all the attendants of a particular flight. However, the same number of attendants contacted the rash on each flight. In addition, a few of those who contracted the rash felt ill, and the union threatened action because the attendants were upset, worried, and believed some malicious force was behind it. Many doctors were called in, but all were in a quandary. Industrial hygienists could not measure anything extraordinary in the cabins. Carry out a K.T. Problem Analysis to see if you can learn the cause of the problem. (*Chemtech*, 13 (11), 655, 1983)

	IS	IS NOT	DISTINCTION
WHAT:	Rash	Other illness	External contact
WHEN:	New planes used	Old planes used	Different materials
WHERE:	Flights over water	Flights over land	Different crew procedures
EXTENT:	Face, hands, arms	Other parts	Something contacting face, hands and arms
	Only some attendants	All attendants	Crew duties

We now look at all the distinctions and see that a) something contacting the arms and face could be causing the rash, b) the rash occurs only on flights over water, and that the use of lifevests are demonstrated on flights over water, and c) the lifevests on the new plane are made of new materials or of a different brand of materials and that usually three flight attendants demonstrated the use of the lifevests. The new life preservers had some material in or on them that was the rash-causing agent!

Oh, Nuts!!! *

The Nuts'n'Bolts Auto Parts Company manufactures and distributes auto parts throughout the United States. Over a period of several months, they have been receiving a large number of complaints about corroded bolts from consumers. Virtually all of the complaints were received between June and August. There were a few complaints during some of the other months, but almost none in January and February.

In addition to its manufacturing plant in Detroit, Nuts'n'Bolts has four major distribution centers in Atlanta, Phoenix, Denver, and Houston, where shipments from Detroit are stored in warehouses. There seems to be a strong geographical pattern to the complaints with respect to where shipments originated. A majority of the complaints came from shipments from the regions in Houston and Atlanta. Virtually no complaints came from the centers at Denver and Phoenix. Sampling indicates that not every part from any given shipment is corroded; only some of the parts, some of the time, from certain geographical locations. Also, due to excellent quality control, virtually no product leaves the plant with any signs of corrosion.

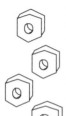

The parts are packaged in cardboard boxes, with cardboard placed in between layers of bolts to act as a shock absorber. A few years ago paper was supplied exclusively by Wolverine Paper, whose plant is located near Lake Superior in Michigan. The newly appointed manager in the Nuts'n'Bolts packaging department noted that Wolverine Paper was overcharging for their product and decided to look into other suppliers.

The best price offered by far (almost 20% cheaper than the next lowest bid) was from Acadia Paper, located in Maine. Research into how Acadia was able to provide such a low bid yielded the following information. The plant was intended to produce high-grade paper, but the water intake for the mill is located in a tidal basin, allowing seawater to enter the processing water supply. Because of this, only low-grade paper can be produced at the plant. In order to get rid of their excess low-grade paper, Acadia began offering packaging paper well below the "market price."

The manager also set up a contract with Badger Paper, whose plant is located near Lake Michigan in Wisconsin. Badger had recently been fined by the Environmental Protection Agency for dumping excessive waste in Lake Michigan. As a result, several changes in Badger's production were made to stay within the EPA's waste limits. This resulted in a decrease in the overall paper quality. In addition to these two suppliers, Nuts'n'Bolts still purchased some of its paper from Wolverine Paper Co. Current prices offered by the paper companies determine which company's paper ends up as packaging material for any batch of product.

- continued -

---

*Based on a true industrial problem. Developed in collaboration with Michael Szachta and Professor Brymer Williams, University of Michigan, 1992.

Oh, Nuts!!!–continued

A confounding factor is that Nuts'n'Bolts has noticed that the quality of the steel to make bolts provided by Heavy Metal has decreased substantially after several key personnel retired. On one occasion Heavy Metal tried to supply materials that arrived exhibiting excessive amounts of corrosion. The president of Nuts'n'Bolts had the shipment sent back to Heavy Metal and threatened to pull the account. After this, there were two other similar incidents of corroded Heavy Metal materials reported. Use a K.T. Problem Analysis to find the cause of the corrosion.

Solution

K.T. Problem Analysis
Problem Statement: What causes the corrosion in the product?

	IS	IS NOT	DISTINCTION	PROBABLE CAUSE
WHAT	Corroded product	Non-corroded product	Corrosion	Salt
WHERE	In Atlanta and Houston	Denver and Phoenix	Type of climate	Effect of humidity on corrosion
	Badger or Acadia Plant	Wolverine Plant	Salt in Acadia No salt in Badger or Wolverine	Tidal basin contamination
WHEN	Summer	Winter	Temperature and humidity	Moist paper
	After new paper companies added	Before new paper companies added	Different paper companies process	Salt in paper
EXTENT	Some of the product	All of the product	Different paper for packaging	Different paper companies
	All bolts in shipment	All bolts in all boxes	Something contacting surface of all bolts	Packaging material

Probable cause: Salt in paper is moving onto parts through water from humid air.

Analysis

In many problems, as in this one, there is an overload of information. A key step in the solution process is to sort out the relevant information from the extraneous facts. For example, in this problem it would be easy to assume that the steel company is responsible for the corroded parts, but this fact does not explain why complaints occur in the summer and why in only certain regions of the country.

- continued -

Oh, Nuts!!!–continued

The only major change recently is the switch in paper companies. But which of the paper companies is causing the problem and how? Wolverine Paper can be eliminated as the cause, since there had never been a problem with them before and no major changes have occurred in their production process. The change in Badger Paper's process could be the cause of the problem, but Acadia Paper seems the most likely candidate, since salts, which are known to cause corrosion, are in the paper already. This fact explains how the salt gets to the parts, but not why only certain geographical locations are affected during certain times of the year.

    Because the problem occurs mostly in the summer time, we might suspect that temperature, in combination with the paper type, would be the cause of the corrosion problem, but the problem doesn't occur in Phoenix, which eliminates temperature as an independent factor. Therefore, we must look at another major change in the affected areas: The arrival of summer brings not only an increase in temperature but an increase in humidity. The paper absorbs the water in the air during humid days, creating a medium through which the salt in the paper is able to contact the metal parts. Once this has occurred, the salts then act to corrode the product.

    As we can see, the use of the Kepner-Tregoe Problem Analysis is beneficial in determining which parts of a problem statement are relevant, thereby facilitating solution of the problem. This problem was adapted from an actual scenario. Unfortunately, because of the problems with their paper, Acadia Paper Co. went out of business.

5.3 DECISION ANALYSIS

    In this section, we will discuss *how to choose the best solution* from a number of alternatives that have been formulated to solve the problem. The K.T. Decision Analysis is a logical algorithm for choosing between different alternatives to find the one that best fulfills all the objectives. The first step is to write a concise *decision statement* about what it is we want to decide and then to use the first four steps discussed in Table 3-1 to gather information.

Choosing a Paint Gun*

A new auto manufacturing plant is to be built and you are asked to choose the electrostatic paint spray gun to be used on the assembly line. The industry standard gun is Paint Right. While experience has shown that Paint Right performs adequately, its manufacturer is located in Europe, making service slow and difficult. In addition, because Paint Right dominates the market, its price is significantly inflated. Two American companies are eager to enter the market with their products: New Spray and Gun Ho.

Decision statement: Choose an electrostatic paint spray gun. The paint guns available are Paint Right, New Spray, and Gun Ho.

*Based on a true industrial problem developed in collaboration with Corinne Falender.

Next we specify the objectives of the decision and divide these objectives into two categories: **musts** and **wants.** The musts are mandatory to achieve a successful solution and they have to be measurable. Next we evaluate each alternative solution against each of the musts. If the alternative solution satisfies all the musts, it is a "go"; if it does not satisfy any one of the musts, it is a "no go," that is, it should not be considered further. In the paint gun example, laboratory experiments showed that Gun Ho could not control the flow of paint at the level required, and thus it was dropped from consideration.

**Set
Criteria**

After learning which alternatives satisfy the musts, we proceed to make a list of the objectives we would want to satisfy. The wants are desirable but not mandatory and give us a comparative picture of how the alternatives perform relative to each other. We list each want, and then assign it a *weight* (1–10) to give us a sense of how important that want is to us. If a want is extremely important, it should be given a weight of 9 or 10. However, if it is only moderately important, such as the "durability of the paint gun," the weight should be a 6 or 7. The next step is to evaluate each alternative against the wants and give it a *rating* (0–10) as to how well it satisfies the want. If the alternative fulfills all possible aspects of a want, it would receive a rating of 10. On the other hand, if it only partially fulfilled the want, it might receive a 4 or 5 rating. For example, in *Choosing a Paint Gun*, the plant personnel are quite experienced at using the current Paint Right spray gun, so it receives a rating of 9 for the experience want. We then multiply the weight of the want by the rating to arrive at a score for the want for that alternative. For the "experience" want, the score is $4 \times 9 = 36$. We do this evaluation for every want and add up the scores for each alternative. The alternative with the highest total score is your tentative first choice.

Assigning weights is indeed a subjective task. However, comparing all the wants two at a time can help to arrive at a consistent assignment of weights. Returning to the paint gun example, there are three pairs of the wants that we can compare: Want 1 with Want 2, Want 2 with Want 3, and Want 3 with Want 1. Let's first compare *ease of service* with *durability* and decide *ease of service* is more important. Next we compare *durability* with *experience* and decide that *durability* is more important. Finally, we compare *experience* with *ease of service*. If we would decide that *experience* is more important than *ease of service*, then we would have an inconsistency that we would have to resolve:

{Ease of Service}	is more important than	{Durability}
{Durability}	is more important than	{Experience}
{Experience}	is more important than	{Ease of Service}

This is like saying, "I like red better than blue, and blue better than green, but I like green better than red." There is an inconsistency that doesn't make sense. The weights and the ordering must be reconsidered in view of the overall picture, so that the final values assigned are consistent.

This situation is obviously easy to avoid with so few wants to consider (three in the paint gun example). However, with a larger list to consider, things become

more difficult, and inconsistencies can arise. There is a story that the U.S. Army was taking a survey regarding the food preferences of enlisted men. The men were provided with a long list of food and asked to indicate their preference for the foods on a scale of 1–10 (10 = like very much, 1 = dislike very much). As a test of the consistency of the information, the Army put several foods on the long list more than one time. Cauliflower was placed on the list twice, once following ice cream, and once following asparagus. The cauliflower entry following ice cream was scored a 3, while the cauliflower entry following asparagus scored a 7. This is clearly inconsistent. The score given to cauliflower was influenced by the foods surrounding it on the list. After ice cream, a real favorite, it scored quite low, while after asparagus, it scored much higher. The warning here is clear: The assessments of the weights for the wants in a decision analysis, while very subjective, must be checked for internal consistency if the decision is to be valid.

While identifying weights and scoring may at first seem somewhat subjective, it is an extremely effective technique for those who can dissociate themselves from their personal biases and arrive at a logical evaluation of each alternative. If the alternative you "feel" should be the proper choice turns out to have a lower score than the tentative first choice, then you should reexamine the weight you have given to each want. Analyze your instincts to better understand which *wants* are really important to you. After this rescoring, if your alternative still scores lower than the others, perhaps your "gut feeling" may be incorrect.

Course of Action: The first step is to break down the important qualities of paint guns, and to decide what you **must** have and what you **want** to have. From your experience and discussions with other paint personnel, you determine that you have two **musts**: 1) adequate control over paint flow rate, and 2) acceptable paint appearance. Also, you identify four **wants**: 1) easy service, 2) low cost, 3) long-term durability, and 4) plant personnel with experience in using the product. Plant records show that Paint Right is able to meet both musts. You then run laboratory experiments with New Spray and Gun Ho to determine whether each of them is also able to meet both musts. The four wants are then weighted, and ratings assigned for each gun that satisfies the musts (as carefully as possible.)

Solution

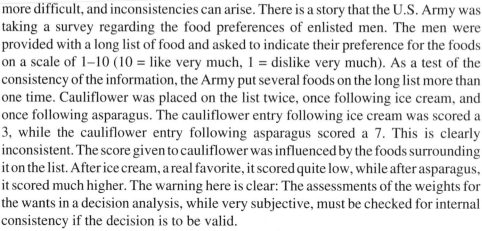

MUSTS		Paint Right		New Spray		Gun Ho
Adequate flow control		Go		Go		No Go
Acceptable appearance		Go		Go		Go
WANTS	Weight	Rating	Score	Rating	Score	
Easy Service	7	2	14	9	63	**NO**
Low cost	4	3	12	7	28	
Durability	6	8	48	6	36	**GO**
Experience	4	9	36	2	8	
Total			110		135	

New Spray was chosen to replace Paint Right.

The last step is to explore the risks associated with each alternative. We take the top-scoring alternatives and make a list of all the things that could possibly go wrong if we were to choose that alternative. We then try to evaluate the *probability* (0–10) that the adverse consequence could occur and the *seriousness* (0–10) of this consequence *if* it were to occur. The product of these two numbers can be thought of as the *threat* to the success of the mission. It is important not to let the numerical scores in the decision table obscure the seriousness of an adverse consequence. In some cases, the second highest scoring alternative may be selected because the adverse consequences of selecting the highest scoring alternatives are too threatening.

Several years ago, a graduating senior from the University of Michigan used K.T. Decision Analysis to help him decide which industrial job offer he should accept. John had a number of constraints that needed to be met. Specifically, his fiancée (now his wife) was also graduating in chemical engineering at the same time and they both wanted to remain reasonably close to their hometown in Michigan. In addition, as a part of a dual-career family, he needed a guarantee that the company would not transfer him. After interviewing with a number of companies, he narrowed his choices to three companies, Dow Corning, ChemaCo, and TrueOil.

The first thing John did was to identify the **musts** that had to be satisfied. These criteria are shown in K.T. Decision Analysis–Job Offer on the next page. Upon evaluating each company to learn if it satisfied all the musts, he found that TrueOil did not satisfy the non-transfer *must*. Consequently, it was eliminated from further consideration. Next all the wants were delineated and a weight assigned to each criterion. The remaining two companies were then evaluated against each want and a total score was obtained for each company. Dow Corning scored 696 points, and ChemaCo. 632 points; the apparent best choice was Dow Corning.

K.T. Decision Analysis—Job Offer

OBJECTIVES MUSTS	Dow Corning	ChemaCo	TrueOil
In Midwest	Midland, MI **GO**	Toledo, OH **GO**	Detroit, MI **GO**
Located w/in 40 miles of spouse's position	Another major company is also in Midland **GO**	Industrialized N. Ohio **GO**	Southeastern Michigan **GO**
Non-transfer policy	Major plant in Midland **GO**	Major plant in Toledo **GO**	**Must Transfer NO GO**

WANTS	Weight	Dow Corning	Rating	Score	ChemaCo	Rating	Score	TrueOil
			Rating/Score			Rating/Score		**NO GO**
Near home town (Traverse City, MI)	8	150 miles	10	80	400 miles	5	40	
Attitude of interviewer	5	Knowledgeable & positive	8	40	Knowledgeable & positive	8	40	
Large company	6	Medium size	6	36	Small size	3	18	
Salary & benefits	9	Good	6	54	Very good	8	72	
Plant safety	10	Good (silicone)	7	70	Mainly oil derivatives (OK)	5	50	
Education assistance program	10	Tuition aid	8	80	Tuition aid	8	80	
Encourage advanced degree	10	Very positive	9	90	Positive	8	80	
Stability of industry	4	Silicone (very good)	9	36	Oil (excellent)	10	40	
Company image	4	Known	5	20	Unknown	3	12	
Type of position	10	Process engineer	9	90	Pilot plant design & operation	10	100	
Advancement policy	7	From within	10	70	From within	10	70	
Return on stockholder investments	3	Excellent (#2 in nation)	10	30	Excellent (#4 in nation)	10	30	
TOTAL				696			632	

However, before making the final decision, the adverse consequences of the first and second choices needed to be evaluated. The results of the adverse consequence analysis are shown in the following table. The adverse consequences analysis ranked both choices in the same order as before, thus the apparent first choice became the final choice.

ADVERSE CONSEQUENCES
Job Offer Analysis

	Probability of Occurrence (P)	Seriousness (S)	PxS
Alternative–Dow Corning			
Wife working in same company	5	7	35
Midland is not very exciting	6	3	18
High rent	4	6	<u>24</u>
Total			77
Alternative–ChemaCo.			
Wife working in same company	3	7	21
Must work nights	6	8	48
High rent	5	6	<u>30</u>
Total			99

Both John and his wife are working at Dow Corning in Midland, Michigan. (Only the names of the other companies have been changed in this real-life example.)

5.3A Cautions

The assigning of weights and scores is indeed very subjective. One could easily abuse this decision-making process by giving higher weights/scores to a predetermined favored project. Such a biased weighting would easily skew the numbers and sabotage the decision-making process. The user is urged to refer to Kepner and Tregoe's book to become aware of certain danger signals that guarantee acceptance of a certain alternative and that blackball all others. This biasing could result from "loaded" want objectives, listing too many unimportant details which obscure the analysis, or a faulty perception of which objectives can guarantee success. Consequently, it is very important to keep an open mind when making your evaluation.

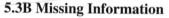

5.3B Missing Information

The most difficult decisions are those where you don't have all the necessary information available upon which to base the decision. Under these conditions it could be helpful after you have prepared a K.T. Decision table to look at the extremes of the missing information and to perform a "What if... ?" analysis. For example, just suppose in the job offer scenario, Dow Corning had not yet decided the type of position John would have with the company. John could assume the best case (his desired position of process engineer) which he would rate at 9.0, and the worst case in his opinion (e.g., traveling sales representative on the road full time) which would give a low rating of 1.0. With this assumption, the total score for Dow Corning would drop to 616 which is now below the score of ChemaCo. We see that this "want" requires a key piece of information and that John must obtain more information from Dow Corning before he can accept their offer. If Dow Corning could not tell John which type of job he would have, they might at least be able to tell him which type of job he might *not* have (e.g., traveling sales representative). If they cannot do the latter, John could have been "forced" to choose ChemCo. On the other hand, if all other factors are positive, John could decide to *take a risk* and choose Dow Corning with the chance he will be able to secure the desired position upon hiring or shortly after being hired.

5.3C Is the Decision Ethical?

While this is an extremely important question, we are going to delay discussion of it until Chapter 7, where we present some thoughts that we hope will help you answer this question.

5.4 POTENTIAL PROBLEM ANALYSIS

Having made our decision, we want to plan to ensure its success. We need to look into the future to learn what could go wrong and make plans to avoid these pitfalls. To aid us in our planning, Kepner and Tregoe have suggested an algorithm that can be applied not only to ensuring the success of our decision but also when analyzing problems involving safety. The K.T. Potential Problem Analysis (PPA) approach can decrease the possibility of a disastrous outcome. As with the other K.T. approach, a table is constructed: The PPA Table delineates the potential problems and suggests possible causes, preventive actions, and contingent actions.

K.T. Potential Problem Analysis			
Potential Problem	Possible Causes	Preventive Action	Contingent Actions
A.	1. 2.		
B.	1. 2.		

In analyzing potential problems, identify how serious each problem would be if it were to occur and how probable it is to occur. Would the problem be fatal to the success of the decision (a must), would it hurt the success of the decision (a want), or would the problem only be annoying? First, we identify all the *potential problems* that could occur and the *consequences* of each occurrence. Be especially alert for potential problems when (1) deadlines are tight, (2) you are trying something new, complex, or unfamiliar, (3) you are trying to assign responsibility, and (4) you are following a critical sequence. Next, list all the *possible causes* that could bring about each problem and develop *preventive actions* for each cause. Finally, develop a *contingent action* (last resort) to be undertaken if your preventive action fails to prevent the problem from occurring. Establish early warning signs to trigger the contingency plan. Do not, however, proceed with contingency plans rather than focusing on preventive actions.

Ragin' Cajun Chicken*

Wes Thompson is a manager of a Burgermeister restaurant, which specializes in fast food hamburgers. He has just been notified by the corporation that a new chicken sandwich, called Ragin' Cajun Chicken, will be introduced into Burgermeister restaurants in two weeks. This surprised Wes because he has never heard anything about the new sandwich from the company or from advertisements. The memo says that plans for a national advertising campaign have unfortunately been delayed until after the introduction of the sandwich.

The memo also says that next week, Wes's restaurant will receive a shipment of 500 Ragin' Cajun Chickens. These are shipped frozen and have a shelf life of three months in the freezer. The notification also stresses the importance of proper handling of the uncooked chicken. In order to prevent cross-contamination by salmonella, the bacteria present in some raw chicken, specially marked tongs will be used to handle only uncooked chicken.

With the shipment of the chicken, Wes's restaurant will receive a new broiler to be used exclusively for the new sandwich. It is important that the broiler operate at least at 380°F to ensure that the chicken will be fully cooked in the five-minute preparation time.

Wes thought that it was very important that the transition run smoothly when Ragin' Cajun Chicken would be added to the menu in two weeks . To prevent any problems, he noted concerns in four areas and constructed the following PPA table:

POTENTIAL PROBLEM	CONSEQUENCE	POSSIBLE CAUSE	PREVENTIVE ACTION	CONTINGENT ACTION

- continued -

*Developed in collaboration with Michael Szachta, University of Michigan, 1993.

Ragin' Cajun Chicken—continued

POTENTIAL PROBLEM	CONSEQUENCE	POSSIBLE CAUSE	PREVENTIVE ACTION	CONTINGENT ACTION
People don't buy sandwich	Restaurant loses money	Customers don't know about sandwich	Make own signs for sandwich	Have cashiers suggest chicken to customers
		Too expensive	Compare unit cost with competition	Run promotional specials
		Food too spicy	Inform customers of mild variety	Run promotional specials
Bacteria in food	Illness, lawsuits	Employees don't handle raw chicken properly	Train employees	Perform periodic inspections
		Improper use of broiler	Train employees	Perform periodic inspections
		Chicken stored too long	Set up dating system	Inspect and discard chicken if necessary
		Freezer not cold enough	Perform temperature checks	Inspect and discard chicken if necessary
Substandard sandwich quality	Customers complain; no return business	Wrong items on sandwich	Have cashiers double-check accuracy	Provide free remade sandwiches for affected customers
		Sandwich sits too long under heat lamps	Mark discard times on sandwiches	Inspect sandwiches before serving
Substandard service quality	Customers complain; no return business	Sandwich preparation takes too long	Always have chicken precooked	Have sandwiches premade

- continued -

Ragin' Cajun Chicken—continued

Of the 10 possible causes for potential problems noted in the table, four occurred. With the strategy in place, possible disaster was averted. Wes noticed the following problems:

• Most customers were unaware of the new menu item. Wes made signs announcing the new sandwich and asked his cashiers to suggest the chicken sandwich (i.e., "Would you care to try our new Ragin' Cajun Chicken today?"). Sales of the sandwich increased dramatically because of this.

• Wes held a special training session for all the employees to explain how critical the proper handling and preparation of the chicken is. Afterwards, Wes also performed periodic inspections and noticed that employees weren't following his instructions (use special tongs and wash hands after handling raw chicken). After a week of inspections, the new operating procedures were being followed by all employees. Fortunately, no cases of food poisoning were reported.

• At the training session, Wes also explained the broiler operating procedures. Once the Ragin' Cajun Chicken was placed on the menu, Wes observed how employees operated the new broiler. Thanks to his observation, an explosion that might have been caused by improper lighting of the broiler was avoided.

• Early on, there were several complaints about improperly made sandwiches. This problem was solved by having cashiers double-check the accuracy of the order before serving the sandwich. This double-checking helped improve the communication between cashiers and cooks, and higher accuracy in sandwich preparation was noticed in all sandwiches.

Lemon-Aid
Buying a Used Car–Not a Lemon

Potential Problem	Possible Causes	Preventive Action	Contingency Plan
1. Buying a car that has improperly aligned front and back wheels.	Car in accident	Pour water on dry pavement and drive through to determine if front and real wheel tracks follow the same path or are several inches off	Don't buy car.
2. Body condition not what it appears to be (concealed body damage).	Car in an accident or body rusted out	Use a magnet along rocker panels, wheel wells, and doors to check for painted plastic filler to which the magnet won't stick. Look under insulation on doors and trunk for signs the car was a different color.	Offer much lower price.
	Car was in a flood, window/trunk leak	Take a deep whiff inside car and trunk. Does it smell moldy? Look for rust in spare tire well.	
3. Car has suspension problems.	Hard use, poor maintenance	Check tire treads for peaks and valleys along the outer edges.	Require suspension be fixed before buying.
4. Leaking fluids.	Poor maintenance	Look under hood and on the ground for signs of leaking fluids.	Require seals be replaced before buying.
5. Odometer not correct.	Tampered with or broken	Check windows and bumpers for decals or signs of removed decals indicating a lot of traveling. Look for excessive wear on accelerator and brake pedals. Check the title.	Offer much lower price.
6. Car ready to fall apart.	Car not maintained during previous ownership	Check fluid levels (oil, coolant, transmission, brake). Check to see if battery terminals are covered with sludge. Check for cheap replacement of oil filters, battery, etc.	Don't buy car.

SUMMARY

Situation Analysis

Problems	Timing (H,M,L)	Trend (H,M,L)	Impact (H,M,L)	Process (PA, DA, or PPA)
1.				
2.				
3.				

Problem Analysis

	IS	IS NOT	Distinction	Probable Cause
What				
Where				
When				
Extent				

Decision Analysis

Alternative:		A		B		C	
Musts	1. 2.	GO GO		GO NO GO		GO GO	
Wants	WT	Rating	Score	Rating	Score	Rating	Score
1. 2.				NO GO ✕			
		Total A=		Total B=		Total C=	

Potential Problem Analysis

Potential Problems	Possible Causes	Preventive Actions	Contingency Plan
A.	1. 2.		
B.	1. 2.		

REFERENCES

1. Kepner, C.H., and B.B. Tregoe, *The Rational Manager*, 2nd ed., Kepner-Tregoe, Inc., Princeton, NJ, 1976.

2. Kepner, C.H., and B.B. Tregoe, *The New Rational Manager*, Princeton Research Press, Princeton, NJ, 1981.

3. Woods, D.L., *A Strategy for Problem Solving*, 3rd ed., Department of Chemical Enginering, McMaster University, Hamilton, Ontario, 1985; *Chem. Eng. Educ.,* p. 132, Summer 1979; *AIChE Symposium Series*, 79 (228), 1983.

FURTHER READING

Kepner, C.H., and B.B. Tregoe, *The New Rational Manager*, Princeton Research Press, Princeton, NJ, 1981. Many more worked examples on the K.T. Strategy.

Keith, Lawrence A., "Report Results Right!," Parts 1 and 2, *Chemtech*, p. 351, June 1991, and p. 486, August 1991. Guidelines to help prevent drawing the wrong conclusions from your data.

EXERCISES

Situation Analysis

1. *The Exxon Valdez.* It is 12:45 AM in the morning, March 24, 1989; you have just been alerted that the Exxon Valdez tanker has run aground on the Bligh Reef and is spilling oil at an enormous rate. By the time you arrive at the spill, 6 million gallons of oil have been lost and the oil slick extends well over a square mile.

   A meeting with the emergency response team is called. At the meeting it is suggested that a second tanker be dispatched to remove the remaining oil from the Exxon Valdez. However, the number of damaged compartments from which oil is leaking is not known at this time and there is concern that if the tanker slips off the reef, it could capsize if the oil is only removed from the compartments on the damaged side.

   The use of chemical dispersants (i.e., soap-like substances) which would break up the oil into drops and cause it to sink is suggested. However, it is not known if there is sufficient chemical available for a spill of this magnitude. The marine biologist at the meeting objected to the use of dispersants, stating that once these chemicals are in the water, they would be taken up by the fish and thus be extremely detrimental to the fishing industry.

   The use of floatable booms to surround and contain the oil also brought about a heated discussion. Because of the spill size, there is not enough boom material even to begin to surround the slick. The Alaskan governor's office says the available material should be used to surround the shore of a small village on a nearby island. The Coast Guard argues that the slick is not moving in that direction and should be used to contain or channel the slick movement in the fjord. The Department of Wildlife says the first priority is the four fisheries that must be protected by the boom or the fishing industry will be depressed for years, perhaps generations to come. A related issue is that millions of fish were scheduled to be released from the fisheries into the oil contaminated fjord two weeks from now. Other suggestions as to where to place the boom material were also put forth at the meeting.

   Carry out a K. T. Situation Analysis on the Exxon Valdez Spill as discussed above.

2. *The Long Commute.* The Adams family of four lives east of Los Angeles in a middle-class community. Tom Adams' commute to work is 45 miles each way to downtown L.A. and he is not in a car or van pool. He has been thinking about changing to a job closer

to his home but has been working for over a year on a project that, if successfully completed, could lead to a major promotion. Unfortunately, there is a major defect in the product which has yet to be located and corrected. Tom must solve the problem in the very near future because the delivery date promised to potential customers is a month away.

Tom's financial security is heavily dependent on this promotion because of rising costs at home. Both children need braces for their teeth, he is in need of a new car (it broke down twice on the freeway this past fall), the house is in need of painting, and there is a water leak in the basement that he has not been able to repair.

Sarah, Tom's wife, a mechanical engineer, has been considering getting a part-time job, but there are no engineering jobs available in the community. Full-time positions are available in Northern L.A., but this would pose major problems with respect to chauffeuring and managing the children. There are a couple of day-care centers in the community, but rumor has it they are very substandard. In addition, last year, their son, Alex was accepted as a new student by the premier piano teacher in the area and there is no public transportation from their home to his studio. Melissa is very sad at the thought of giving up her YMCA swimming team and her girl scout troop, which both meet after school.

Carry out a K. T. Situation Analysis on the Adams family's predicaments.

3. Make up a situation similar to Exercises 1 and 2 and carry out a K.T. Situation Analysis.

Problem Analysis

4. *Off-Color Tooth Paste*. After Crest™ tooth paste had been on the market for some time, Procter & Gamble, its manufacturer, decided to offer a mint-flavored version in addition to the original, wintergreen-flavored product.

In the course of developing the new mint-flavored product, a test batch of mint product was produced by the same pilot unit used to produce wintergreen-flavored product. The pilot equipment uses a tank and impeller device to mix the mint flavor essence with the rest of the ingredients to form the finished product (which is a very viscous solution).

Some of the pilot plant product was packed into the familiar collapsible tubes for further testing. Tubes used in testing the mint flavor were identical to those used for the wintergreen-flavored product. In the packing operation, toothpaste is pumped through lines into the as-yet unsealed ends of brand new tubes. After filling, the open tube ends are heat-sealed. The packing operation is illustrated in the figure.

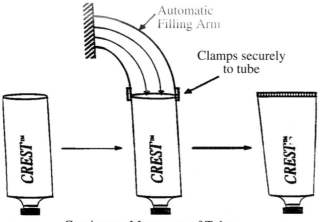

Continuous Movement of Tubes

To assess storage stability, some of the filled tubes were randomly separated into several groups and each group was stored in a constant temperature room. Storage temperatures varied from 40°F to 120°F. Early sampling of the stored product showed nothing unusual. However, several months into the test, a technician preparing to test the product from one of the stored tubes noted that the first 1/4" of paste squeezed onto a toothbrush was off-color. The rest of the product in the tube met the color specification. Nothing like this had ever been seen with the original formula.

Further testing showed that one had to squeeze more product out of those tubes that had been stored at higher temperatures and/or stored for longer times before a product that met color specs would exit the mouth of the tooth paste tube. Tubes stored for a period of time at 40°F contained no off-color product while tubes stored for the same length of time at higher temperatures produced off-color paste. The only exception to these results was a single tube, stored above 40°F. A leakage of off-color product was found around the base of the cap on this tube, but the product inside the tube met color specs. While other tests showed the off-color product to be safe and effective in cleaning teeth, consumers clearly would not accept a color change in a product expected to have the same color from the first squeeze to the last. Moreover, such a change could have been an early warning of more serious problems to come. This phenomenon had to be understood and eliminated before the new flavor could be marketed.

Accordingly, various possible remedies were tested: caps and tubes made of different materials, different mixing methods, etc. None of these had any effect on the off-color problem. All raw materials, including the new mint flavor essence, were checked and found to meet specs. A subsequent batch of the wintergreen product was made and tested for storage stability, and as usual, no off-color problems occurred.

Carry out a K.T. Problem Analysis to learn the cause of the off-color tooth paste.

5. *Chocolate Covered Bacteria.* Chocolate butter paste is the primary ingredient used by a number of major bakeries for a wide variety of pastries. The paste is a very viscous liquid that is manufactured by *Cocomaker Industries* in a major populous city in the midwest. *Cocomaker* supplies customers as close as Dolton, and as far away as Chicago, which is a long drive. The paste flows from the production line into five-gallon drums, which are placed immediately into refrigerated trucks for shipment to the respective customers. Until February, all the trucks were the same size and the drums were stacked in rows three drums wide, four drums high, and eight drums deep. However, now two rather small customers each requiring 20 drums per day were added in the Chicago area, which, along with an increased order by the Chicago customer Hoyne, necessitated the purchase of a larger truck. The new truck could fit five drums across, four drums high, and eight drums deep. The truck would stop at the two smaller additions, Bell Bakery and Clissold Bakery, just before and just after stopping at Hoyne Industrial Bakers in Chicago proper.

With the increased market in the Chicago area, *Cocomaker* is running at close to maximum capacity. Because the ingredients of the paste are mixed by static mixers, the pumps are currently operating at their maximum capacity and the plant is operating 20 hours per day. In November, *Cocomaker* was successful in luring two nearby customers, Damon Bakery and Oakley Bakery, away from one of its competitors. By increasing plant operation to 24 hours per day, all orders could be filled.

As the Christmas season approaches, the usual seasonal demand for the chocolate butter paste poses a problem of meeting demands not encountered in previous years. It was decided that if the processing temperature were increased by 20 degrees, the paste would be sufficiently less viscous, and production demands could be met with the current pump limitations. However, the increased capacity began to generate problems as Christmas approached. The pumps began failing on a regular basis; a strike at the supplier of the shipping containers caused *Cocomaker* to buy from a new container supplier, which claimed to carry only sturdier containers at a 10% increase in price; the safety officer had an emergency appendectomy; and most troubling, Hoyne Industrial Bakeries have been calling about an unacceptable bacteria count in shipments for the last five days. As a result, buyers of their product may have been getting ill. An immediate check of the bacteria levels show that they are at the same acceptable levels they have always been when leaving *Cocomaker*. You call Mr. Hoyne and tell him that the plant levels show that the paste is within bacteria specifications. Two days later you receive a call from Hoyne saying that they hired an independent firm and they reported the bacteria levels are well above an acceptable level. You call Damon, Bell, Clissold, and Oakley bakeries and ask them to check their bacteria count; they report back that everything is within specifications most often reported. A spot check of other customers shows no problems. You receive a call form Hoyne saying they are starting legal and governmental actions to close you down.

Carry out a K.T. Problem Analysis to learn the cause of the problem.

6. *Toxic Water* . Sparkling mineral water is the primary product of Bubbles, Inc., based in France, which serves three major markets in Europe, North America, and Australia. The water is collected from a natural spring and filtered through a parallel array of three filter units, each containing two charcoal filters. The filtration process is needed to remove trace amounts of naturally occurring contaminants. The filtered water is stored in separate tank farms, one for each market, until it is transported by tanker truck to one of the three bottling plants that serve the company's markets.

When the water arrives at the bottling plant, it is temporarily placed in 3500 m^3 storage tanks until it can be carbonated to provide the effervescence that is the trademark of the producer. Some of the water is also flavored with lemon, cherry, or raspberry additives. The sparkling water is then packaged in a variety of bottle sizes and materials from 10 oz. glass bottles to 1 liter plastic bottles. The European market receives its shipments directly by truck, usually within three days. Product bound for North America or Australia is shipped first by truck to the waterfront and then by freighters to their overseas destinations.

Business has been good for the last several months, with the North American and European markets demanding as much sparkling water as can be produced. This situation has required additional plastic bottle suppliers to keep up with the increased demand. It has also forced regularly scheduled maintenance for the Australian and North American markets to be delayed and rescheduled because of the high demand for the product. There is also, of course, a larger demand placed on the spring that supplies the mineral water for the process.

Unfortunately, all news is not good for Bubbles, Inc. The bottling plant for the Australian market is currently several weeks behind schedule due to a shipment lost at sea. This catastrophe has required that water from the company's reserve springs, which are located many miles from the bottling plant, be used to augment the water supplied by the regular spring so that the bottling plant can operate at an even higher level of production. The availability of water from the reserve springs is hindered by their remoteness, but the water from these springs does not require filtration. In addition, contract negotiations are going badly and it appears there will be a strike at all of the bottling plants. Recent weather forecasts indicate that relief from the drought that has already lasted three months is not likely.

Worst of all, the North American and Australian markets are complaining that all shipments of the sparkling water in the last six weeks have contained benzene in unacceptably high concentrations. You know that benzene is often used as an industrial solvent but is also found naturally. A quick survey of the bottling plant managers shows that the North American-bound products currently packaged and awaiting shipment have benzene concentrations in excess of acceptable concentrations. However, the managers of the bottling plants that service the Australian and European markets report that no significant level of benzene was detected in the bottles currently stored. The North American and Australian markets have already begun recalling the product, with the European market pressuring for a quick solution and threatening to recall products as a precautionary measure. (Adapted from *Chemtech,* "When the Bubble Burst," p. 74, Feb. 1992)

    Carry out a K.T. Problem Analysis to learn the cause of the problem.

7. Currently there are many platforms in the Gulf of Mexico that collect the oil from a number of wells and then pump it through a single pipeline from the platform to shore. Most of these wells have always been quite productive and consequently the oil flows through the pipeline lying on the ocean floor at a reasonable rate. When the oil comes out of the wells it is at temperatures of approximately 145°F, and by the time the oil reaches shore the temperature in most pipelines is around 90°F. The temperature of the water on the ocean floor for the majority of the platforms within two miles of shore is approximately 42°F. However, the water depth increases as you move away from shore and the temperature of the water on the ocean floor decreases.

    Recently two new platforms (A and B) were erected in the Gulf Coast farther out from shore than the others. About a year and a half after they both came on stream a disaster occurred on Platform A. No oil was able to be pumped to shore through the pipeline from Platform A. However, Platform B was operating without any problems. When the crude compostion at the well head was analyzed it was found to be the same weight percent composition (e.g., asphaltenes, waxes, gas) as that found in all the well heads on all other platforms. The only difference between Platform A and Platform B was that the production rate of Platform A was much less than that of Platform B. However, the produciton rate from Platform A was still greater than many of the platforms near the shore line.

    Carry out a K.T. Problem Analysis to learn the reason for the plugging of the pipeline.

Decision Analysis

8. *Buying a Car.* You have decided you can spend up to $12,000 to buy a new car. Prepare a *K.T. Decision Analysis* table to decide which car to buy. Use your local newspaper to collect information about the various models, pricing, and options and then decide on your *musts* (e.g., air bag) and your *wants* (e.g., quadraphonic stereo, CD player). How would your decision be affected if you could spend only $9,000? What about $18,000?

9. *Choosing an Elective.* You need one more three hour nontechnical course to fulfill your degree requirements. Upon reviewing the course offerings, and the time you have available, you note the following options:

- Music 101   Music Appreciation–2 hours
- Art 101   Art Appreciation–3 hours
- History 201   U.S. History; Civil War to Present–3 hours
- Art 203   Photography–3 hours
- Geology 101   Introductory to Geology–3 hours
- Music 205   Piano Performance–2 hours

Music 101 involves a significant amount of time outside of class listening to classical music. The student reaction to the class has been mixed; some students learned what to listen for in a symphony, while others did not. The teacher for this class is knowledgeable but boring.

Art 101 has the students learn the names of the great masters and how to recognize their works. The lecturer is extremely boring and you must go to class to see the slides of the great art works. While the course write-up looks good, it misses the mark in developing a real appreciation of art. However, it is quite easy to get a relatively good course grade.

History 201 has an outstanding lecturer that makes history come alive. However, the lecturer is a hard grader and C is certainly the median grade. In addition, the outside reading and homework are enormous. While some students say the work load is equivalent to a five-hour course, most all say they learned a great deal from the course and plan to continue the interest in history they developed during this course.

Art 203 teaches the fundamentals of photography. However, equipment and film for the course are quite expensive. Most of the time spent on the course is outside of class looking for artistic shots. The instructor is very demanding and bases his grade on artistic ability. Some students say that no matter how hard you work, if you don't develop a "photographic eye" you might not pass the course.

Geology 101 has a moderately interesting lecturer and there is a normal level of homework assignments. There are two major out-of-town field trips that will require you to miss a total of one week of class during the term. The average grade is B and there is nothing conceptually difficult nor memorable about the course.

Music 205 requires you pass a tryout to be admitted to the class. While you only spend 1/2 hour a week with your professor, many, many hours of practice are required. You must have significant talent to get a C or better.

Prepare a K.T. Decision Analysis Table to decide which course to enroll in.

10. *The Centralia Mine Fire.* Centralia, Pennsylvania, a small community situated in the Appalachian mountain range, was once a prosperous coal mining town. In 1962, in preparation for the approaching Memorial Day parade, the landfill of Centralia was set afire in order to eliminate odors, paper buildup, and rats. Unfortunately, the fire burned down into the passageways of the abandoned mine shafts under the town. Although repeated efforts were made to stop the blaze, the fire could not be put out. By 1980, after burning for 18 years, the fire had grown in size to nearly 200 acres, with no end in sight.

   Mine fires are especially difficult situations because they are far below the surface of the earth, burn very hot (between 400°F and 1000°F), and give off both toxic and explosive gases, as well as large volumes of steam when the heat reaches the water table. Anthracite coal regions have very porous rock, and consequently, a significant amount of combustion gas can diffuse directly up through the ground and into people's homes. Subsidence, or shifting of the earth, is another serious condition arising from the fire. When the coal pillars supporting the ceilings of mines' passageways burn, large sections of earth may suddenly drop 20 or 30 feet into the ground.

   Clearly, the Centralia mine fire has very serious surface impact and must be dealt with effectively. Several solutions to the mine fire are described below. Perform a K. T. Decision Analysis to decide which is the most effective method to deal with the fire. Consider such issues as cost, relocation of the town of Centralia, and potential success of extinguishing the fire.

Solution Options

   1. *Completely excavate the fire site*–Strip mine the entire site to a depth of 435 feet, digging up all land in the fire's impact zone. This would require partial dismantling of Centralia and nearby Byrnesville for upwards of ten years, but available reclamation techniques could restore the countryside after this time. This method guarantees complete extinction at a cost of $200 million. This cost includes relocation of families, as well as the restorative process.

2. *Build cut-off trenches*–Dig a trench to a depth of 435 feet, then fill with a clay-based noncombustible material. Behind the trench, the fire burns unchecked, but is contained by the barrier. Cost of implementation would be about $15 million per 1000 feet of trench, and total containment of the fire would require approximately 7000 feet of trench. Additionally, partial relocation of Centralia would be required for three years, costing about $5 million.

3. *Flood the mines*–Pump 200 million gallons of water per year into the mine at a cost of $2 million annually for 20 years to extinguish the fire. Relocation of the townspeople is not necessary, but subsidence and steam output should be considered, as well as the environmental impact and trade-offs of the large quantities of acidic water produced by this technique.

4. *Seal mine entrances to suffocate fire*–Encase the entire area in concrete to seal all mine entrances, then allow the fire to suffocate due to lack of air. This would require short-term relocation of the towns and outlying areas, and suffocation itself would probably take a few years owing to the large amount of air in the shafts and in the ground. Although this method has never been attempted, the cost is estimated to be about $100 million.

5. *Use fire extinguishing agents*–Pump halons (gaseous fluorobromocarbons) into the mines to extinguish the blaze. The cost for this method would be on the order of $100 million. Relocation may be necessary.

6. *Do nothing* –Arrange a federally funded relocation of the entire area and allow the fire to burn unchecked. Approximately $50 million would be required to relocate the town.

(This problem developed by Greg Bennethum, A. Craig Bushman, Stephen George, and Pablo Hendler, University of Michigan, 1990)

11. You need energy for an upcoming sports competition. You have the following candy bars available to choose from: Snickers™, MilkyWay™, Mars Bar™, Heath Bar™, Granola Bar. Which do you choose? Prepare a K.T. Analysis Table.

12. Prepare a K.T. Decision Analysis Table on selecting an apartment to move into next term (year). Consult your local newspaper to learn of the alternatives available.

Potential Problem Analysis

13. *Sandy Beach.* There was a minor oil spill on a small sandy resort beach. The CEO of the company causing the beach shoreline to be soiled with oil said: "Spare no expense, use the most costly method, steam cleaning, to remove the oil from the sand." (Adapted from *Chemtech*, August 1991, p. 481)

    Carry out a K.T. Potential Problem Analysis on the direction given by the CEO.

14. *Laboratory Safety.* The procedure in a chemistry laboratory experiment called for the students to prepare a $1.0\,dm^3$ aqueous solution of 30 g of sodium hydroxide. By mistake, the student used 30 g of sodium hydride dispersion which reacted violently with water, evolving heat and hydrogen gas which caught fire. The sodium hydride, which was available for a subsequent experiment, was a commercial product. The container bore a warning of the hazard of contact with water, but this warning was not visible from the side showing the name of the compound. (Adapted from *ICE Prevention Bulletin*, 102, p. 7, Dec. 1991)

    Carry out a Potential Problem Analysis that, if followed, would have prevented this accident.

15. *Safety in the Plant.* A reactor approximately 6 feet in diameter and 20 feet high in an ammonia plant had to be shut down to repair a malfunctioning nozzle. The nozzle could be repaired only by having a welder climb inside the reactor to carry out the repair. During welding, the oxygen concentration was regularly monitored. Four hours after the welding was completed, a technician entered the reactor to take pictures of the weld. The next day he was found dead in the reactor. (Adapted from *ICE Prevention Bulletin*, 102, p. 27, Dec. 1991)

    Prepare a Potential Problem Analysis Table that could have prevented this accident.

16. *New Chicken Sandwich.* Burgermeister has been serving fast food hamburgers for over 20 years. To keep pace with the changing times and tastes, Burgermeister has been experimenting with new products in order to attract potential customers. Product development has recently designed a new Cajun chicken sandwich to be called Ragin' Cajun Chicken (see example on page 105). The developers have spent almost nine months perfecting the recipe for this new product.

    One of the developers got the idea for a new product while in New Orleans during last year's Mardi Gras. Product Development has suggested that the sandwich be placed on Burgermeister's menu immediately, in order to coincide with this year's Mardi Gras festivities. A majority of the time spent developing the Ragin' Cajun Chicken sandwich was dedicated to producing an acceptable sauce. Every recipe was tasted by the developers, who found early recipes for sauces to be too spicy. Finally, they agreed on the seventy-eighth recipe for sauce (Formula 78) as the best choice.

    After converging on a sauce, the Development Team focused on preparation aspects of the new sandwich. Several tests confirmed that the existing equipment in Burgermeister restaurants could not be used to prepare Ragin' Cajun Chicken. Instead, a new broiler would have to be installed in each of the 11,000 Burgermeister restaurants, at a cost of over $3,000 per unit. The new broiler would keep the chicken moist while cooking it, as well as killing any salmonella, the bacteria prevalent in chicken.

    While testing cooking techniques for the new broiler, one of developers became very ill. A trip to the hospital showed that the developer had food poisoning from salmonella. Tests determined that the source of the bacteria was a set of tongs that the developer used to handle both the raw and the cooked chicken.

    Next, the Development Team decided how the sandwich would be prepared. When the Ragin' Cajun Chicken sandwich was prepared using buns currently used for other Burgermeister sandwiches, the sandwich received a very low taste rating. After experimenting, researchers found that a Kaiser roll best complemented the sandwich. Early cost estimates showed that Kaiser rolls will cost twice as much as the buns used currently for hamburgers, and are fresh half as long.

    You are an executive in charge of product development for Burgermeister. Based on the information above, perform a Potential Problem Analysis, considering what could go wrong with the introduction of this new sandwich.

    (Developed in collaboration with Mike Szachta, The University of Michigan, 1992)

17. *Choices.* Carry out a Potential Problem Analysis for
    a) A surprise birthday party.
    b) A camping trip in the mountains.
    c) The transportation of a giraffe from the Detroit Zoo to the Los Angeles Zoo.
    d) An upcoming laboratory experiment.
    e) The transport of nuclear waste from the reactor to the disposal site.

6 *IMPLEMENTING THE SOLUTION*

Many people get stalled in the problem-solving process because they analyze things to death and never get around to acting. In this chapter we will present a number of techniques that will facilitate the *implementation* process. Figure 6-1 identifies the phases of the implementation process.

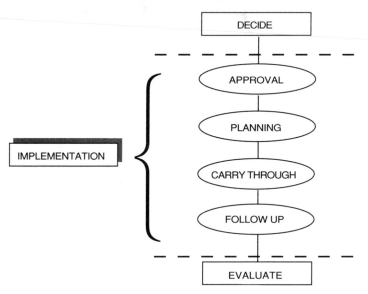

Figure 6-1. Implementing the Solution

Harshberger's Law: *"If you don't know how to do it, but it is important to do it, go ahead and do it anyway."*

Each of these phases will be discussed in this chapter.

6.1 APPROVAL

In some situations, the first step in the implementation is to get approval from your organization to proceed with the chosen solution. Many times it may be necessary to *sell your ideas* so that your organization will provide the necessary resources for you to successfully complete your project. This process may include the preparation of a document or presentation describing 1) what you want to do, 2) why you want to do it, 3) how you are going to do it, 4) how your project will greatly benefit the organization and/or others. The following short checklist can help sell your ideas.

- Avoid technical jargon–keep the presentation clear and to the point.
- Make the presentation in a logical and orderly manner.
- Be concise; avoid unnecessary minute details.
- Anticipate questions and be prepared to respond to them.
- Be enthusiastic about your ideas or nobody else will be.

6.2 PLANNING

Now that you have the resources for the process, it is time to plan what to do, what order to do it in, and when to do it. The most important aspect of *implementing* is the planning stage. Here we look at allocations of time and resources, anticipate bottlenecks, identify milestones in the project, and identify and *sketch the pathway through to the finished solution*. After examining the various parts of the solution which are to be implemented, criteria are needed to decide which part to work on first. A modified K.T. Situation Analysis will help to identify the critical elements of the solution and to prioritize them in order to prepare a meaningful plan. Gantt Charts, Development Charts, Budgets and Critical Path Management[1] will be used to effectively allocate our time and resources. Finally, we proceed to identify what could go wrong and devise ways to prevent these roadblocks from occurring (K.T. Potential Problem Analysis). Market surveys are often used as a part of K.T. PPA to anticipate the possible success or failure of a product or process (e.g., chicken ripple chip ice cream).

To aid us in our planning, we draw on two topics previously discussed: K.T. Situation Analysis (Chapter 5), and K.T. Potential Problem Analysis (Chapter 5).

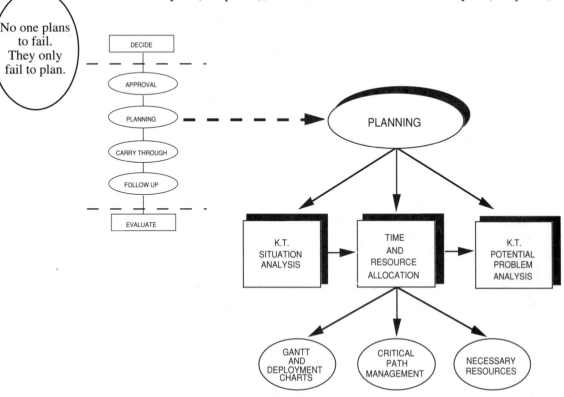

Figure 6-2. Components of the Planning Process

6.2A Allocation of Time and Resources

Having been presented with a problem, situation, or opportunity, we need to allocate our time and resources to the various steps to bring about a successful solution. The Gantt and Deployment charts, Critical Path Management, along with budgeting of personnel and money, can be used to arrive at an efficient and effective allocation. Additionally, a popular tool for scheduling daily activities is the personal organizer (e.g., the Franklin Day Planner™) used by executives and students alike to keep track of important appointments and commitments.

6.2A.1 Gantt Chart

One of the most common ways used to allocate specific blocks of time to the various tasks in a project is the *Gantt Chart*. A Gantt Chart is a bar graph that shows when a specific task is to begin and how long it will take to complete. For the sake of discussion, suppose we have a time constraint of one year to solve the problem and we are to allocate time to each of the five building blocks of the problem-solving process. January (J), February (F), and March (M) will be spent working the problem definition and mid-March to May will be devoted to generating solutions. Note that we have suggested that time be taken to evaluate our progress at four different points along the way to check that all criteria are fulfilled: 1) after completion of the definition of the problem, 2) after deciding the course of action, 3) during the course of action, and 4) at the end of the project.

TABLE 6-1: The Gantt Chart

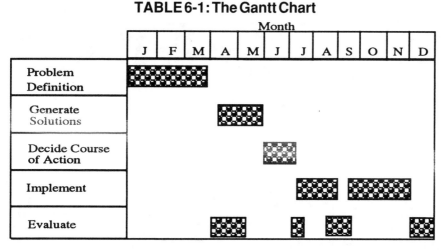

> *"If you don't know where you're going you'll probably end up somewhere else."*
> -Yogi Berra
> N. Y. Yankees

Note that at least 25% of the time has been devoted to the problem definition phase. Many, if not most, of the consequences of incorrectly defined problems discussed in the first chapter would not have occurred if more time had been spent defining the problem rather than hurrying to start a solution. Most experts agree that the project is half completed once the real problem is defined, written down, and communicated.

We now return to the heat exchanger problem discussed in Chapter 3 in which it was learned that the **real problem** was to remove the scale from the heat exchanger rather than designing and building a larger heat exchanger. Typically there can be scale on the inside and on the outside of the heat exchanger tubes. If we have an organic liquid on the outside of the tubes and water inside the tubes, the scales on either side of the tube will be different in nature. The organic scales are tar-like and the water scales are usually mineral salts.

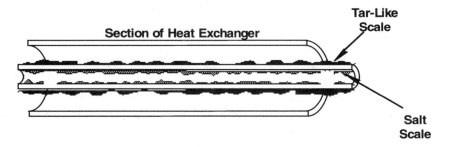

As shown in the Gantt Chart below, a day and a half (all day Monday and Tuesday morning) are devoted to disassembling the heat exchanger. Two and a half days, Tuesday noon through Thursday, are allotted to remove the scale and get the lab results. Each of the remaining tasks in the process is scheduled in a similar manner.

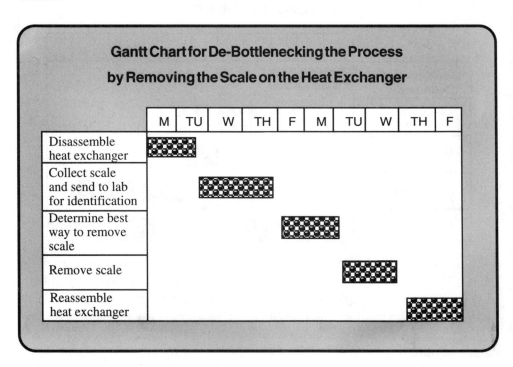

6.2B Coordination and Deployment

Most often groups of individuals will work together as a team to solve a problem. Under these circumstances, coordination among various team members is imperative to achieving an efficient solution in the time allotted. The use of a *Deployment Chart* will help guide the team through the solution by assigning different team members either major or minor responsibilities to each of the tasks.

For example, let's again consider the example of cleaning the scale (fouling) from the heat exchanger in order for it to operate more efficiently. Cesar and Stan will disassemble and reassemble the equipment. Linda will analyze the scale to determine the type and amount. Sheila will help Linda with the analysis and will also be the one responsible for seeing that the scale is properly removed. The remaining tasks and assignments are shown in the Deployment Chart for cleaning the heat exchanger on page 124.

An example of the use of a Gantt Chart that most people can relate to is the preparation of a Thanksgiving turkey dinner.

Thanksgiving Dinner

Our extended family consists of 25 people who will be attending our celebration. Let's consider our dinner menu:

		Time requirements
Main Course:	Roasted turkey (of course) with dressing (stuffing)	clean (1/2 hr), stuff (1/2 hr), cook (7 hrs at 350°F), cool and slice (1 hr)
Vegetable:	Green beans with mushroom sauce	prep. time (30 min), microwave (30 min)
Potato:	Sweet potato casserole	prep. time (30 min), cook (2 hrs at 350°F)
Sauce:	Jellied cranberry sauce	open can, slice, serve
Dessert:	Pumpkin Pie	prep. time (3/4 hrs), cook (1 hr at 425°F)
Beverages:	Coffee, Tea, Milk, Water, White wine	prep. time (minimal)

Clearly, the successful preparation and serving of the Thanksgiving meal will require substantial planning and coordination by the cook. A Gantt Chart will help us organize our time and resources (stove, microwave, etc.). We'd like to sit down to eat at 4 PM. The longest "lead-time" item is the turkey, which requires 9 hours to prepare. Thus, using the turkey as a yardstick, we must begin our preparations by at least 7 AM (a long day for the cook!). Let's try to fit this into a Gantt Chart for the preparation of the entire meal.

- continued -

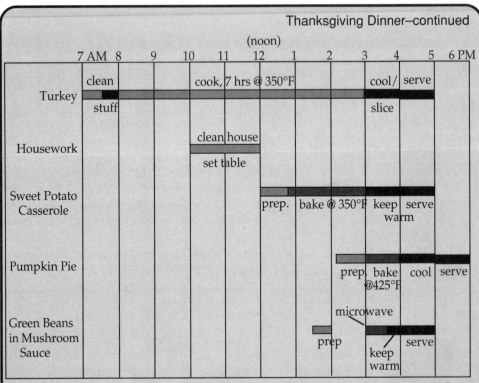

Thanksgiving Dinner–continued

The Gantt Chart shows the hectic nature of the Thankgiving dinner preparations clearly. Any time conflicts should be apparent from the chart.

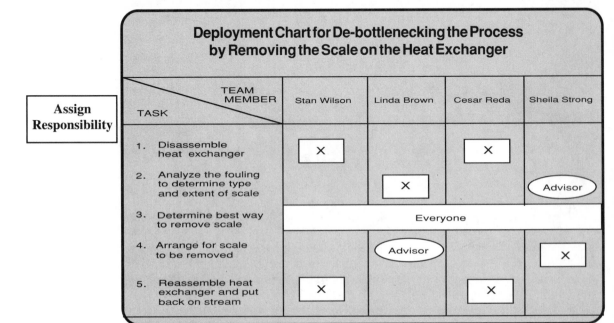

Deployment Chart for De-bottlenecking the Process by Removing the Scale on the Heat Exchanger

Assign Responsibility

TASK / TEAM MEMBER	Stan Wilson	Linda Brown	Cesar Reda	Sheila Strong
1. Disassemble heat exchanger	X		X	
2. Analyze the fouling to determine type and extent of scale		X		Advisor
3. Determine best way to remove scale	Everyone			
4. Arrange for scale to be removed		Advisor		X
5. Reassemble heat exchanger and put back on stream	X		X	

6.2C Critical Path

We use critical path management to identify the critical points in the process. These critical points are readily identified by determining which tasks will cause a substantial delay in the implementation of the solution if the schedule is not met.

As an example of critical path management, let us return to the Cleaning the Heat Exchanger example. The organic scales are typically removed by dissolving them with an appropriate solvent, while the water (mineral) scales are removed with high-pressure water jets. Removing the organic tar scale by soaking in a solvent is a much slower process than high-pressure jet cleaning. Thus, the tasks associated with cleaning the organic tar is the "critical path" for keeping the project on schedule. If any of the tasks associated with this removal are delayed, the overall project will be delayed.

The figure below shows a critical path time line diagram. The critical path is indicated by the heavy black lines. Particular attention must be given to the tasks on this path so that they stay on schedule and the overall project is not delayed.

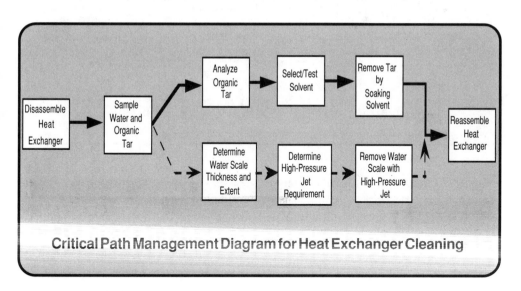

Critical Path Management Diagram for Heat Exchanger Cleaning

Another example of the use of critical path management is for the preparation of the annual Thanksgiving dinner. For the meal to be served on time (and to keep the guests happy) there are several critical steps in the preparation that must be completed in a timely manner or substantial delays will result. In the figure below, for example, the bold lines and boxes indicate the critical path. It refers to items that require a fair amount of time to complete. If the schedule "slips," the meal will be delayed. For example, the turkey requires approximately eight hours to clean, stuff, and cook. If the preparation of the turkey is delayed, chances are the serving time will

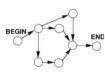

be delayed too. The sweet potato casserole also requires a long time to complete (more than 3.5 hours), so it is critical that it is completed on schedule, or the meal will be delayed. Noncritical path items, such as setting the table, cleaning the house, and preparing the pie for baking, can be done as time permits after the critical items are completed.

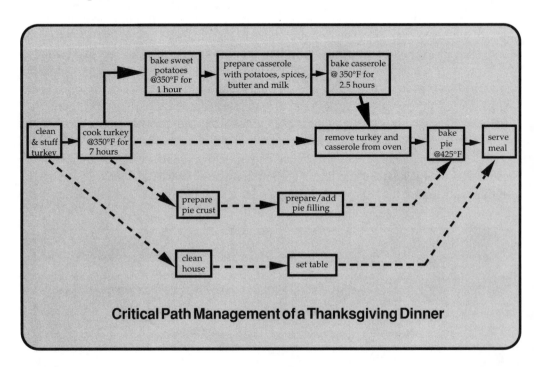

Critical Path Management of a Thanksgiving Dinner

A word of caution is in order here. If noncritical path items are completed too slowly, they can become critical path items. Thus, making a critical path diagram is a dynamic process. The diagram should be continually updated as tasks are completed, viewing the project as a whole.

6.2D Necessary Resources

We must also estimate the resources necessary to complete the project. The resources usually fall into five categories: available personnel, equipment, travel, supplies, and overhead. The contingency funds are to cover unexpected expenses (e.g., extra lumber, more expensive dishwasher). At the start of the project, it is usually important to obtain an estimate of the total cost by preparing a budget similar to that shown on the next page.

Proposed Budget
BUDGET

Personnel	Months/Rate	Cost	% of Total
Tom Smith, Project Director	6 mos @ $5000/mo	$30,000	
Bill Wade, Engineer	12 mos @ $3500/mo	42,000	
Secretary	2 mos @ $2250/mo	4,500	
Subtotal - Salaries		76,500	44%
Fringe Benefits 24% of Salaries		18,360	10%
Total Salary and Benefits		$94,860	
Equipment			
Fabrication in machine shop			
Parts and Labor		$10,000	6%
Travel			
Attend Professional Meeting		$1,140	0.7%
Supplies			
Chemicals, etc.		$4,000	2.3%
Subtotal		$110,000	
Overhead 58%		$63,800	37%
TOTAL BUDGET		$173,800	100%

6.3 CARRY THROUGH

The Carry Through Phase is an essential step in a successful solution process. In this phase, the various people involved in the problem-solving process act upon the plans they have formulated. Here they may carry out a design, fabricate a product, carry out experiments, make calculations, prepare a report, cook a dinner, go on an activity, etc. There are some instances in which the *Implementation* phase and the *Deciding the Course of Action* phase are intertwined. For example, it may be necessary to collect experimental or other data (implement a plan) before the right decision can be made. Great care should be taken with this phase. All the planning in the world will not save a poor job of carrying through the chosen solution. Table 6-2 provides a check list of things to monitor in the Carry Through Phase.

Monitor Progress

TABLE 6-2: Carry Through Check List

- Find the limits of your solution by making different simple models or assumptions that would clearly both
  - overestimate the answer, and
  - underestimate the answer.

- Make an educated guess of what your solution will look like.

- Construct a quick test or experiment to see if the solution you have decided upon will work under the simplest conditions.

- Continue to learn as much as you can about the solution you have chosen. Read the literature and talk to your colleagues.

- Continue to challenge and/or validate the assumptions of the chosen solution. Make sure no physical laws are violated.

- Plan your computer experiments (i.e., simulations) as carefully as you would plan your experiments in the laboratory.

Revealing the Solution

A procedure for *carrying through* a solution that has many facets or components is one that is adapted from Bloom's Taxonomy. In this procedure, the activities are arranged from the most difficult (synthesis) to the easiest (comprehension). The description of these activities carried out is as follows: Evaluation, Synthesis, Analysis, Application, Comprehension, and Knowledge. The *Carry Through Process* adapted from Bloom's Taxonomy is shown at the left.

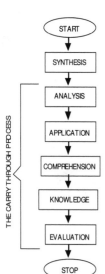

Evaluation: Evaluation is an *ongoing* process throughout the entire problem-solving process. Qualitative and quantitative judgments about the extent to which the materials and methods satisfy the external and internal criteria should be made.

Synthesis: This activity is the putting together of parts to form a new whole. Synthesis enters into problem solving in many ways. Given a fuzzy situation, synthesis is the ability to formulate (synthesize) a problem statement and/or the ability to propose a method of testing hypotheses. At the end of this activity we have *defined* the problem, *generated* a number of potential solutions and decided on which solution to implement.

Once the various parts are synthesized, each part (problem) now uses the intellectual skill described in *analysis* to continue toward the complete solution.

Analysis: This activity is the process of breaking the problem into parts such that a hierarchy of subproblems or ideas is made clear, and the relationship among these ideas is made explicit. In analysis, one identifies missing, redundant, and contradictory information.

Once the analysis of a problem is completed, the various subproblems are then reduced to problems requiring the use of *application skills*.

Application: This activity recognizes *which set* of principles, ideas, rules, equations, or methods should be applied, given all the pertinent data.

Once the principle, law, or equation is identified, the necessary knowledge is recalled, and the problem is solved as if it were a comprehension problem.

Comprehension: This activity involves understanding, manipulation, and/ or extrapolation of the knowledge (i.e., principle, equation) we identified in the *Application Step* to solve a given problem. That is, given a familiar piece of information, such as a scientific principle, can the problem be solved by recalling the appropriate information and using it in conjunction with manipulation, translation, or interpretation of the equation or scientific principle?

Knowledge: Knowledge is remembering previously learned material. It is used in each step of Bloom's method of unraveling. Here we ask, "Can the problem be solved simply by defining terms and by recalling specific facts, trends, criteria, sequences, or procedures?"

The main advantage of using Bloom's Method is that it allows us to unravel the solution. That is, completion of each step (e.g., analysis) uncovers the next step to be worked on (e.g., application). Of course, additional knowledge must be injected into each step along the way.

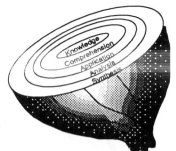

It's Like Peeling an Onion!

As an example of this "unraveling" procedure, let's apply the adaptation of Bloom's Method to unravel a problem and see how it works. The Carry Through process actually begins after we have defined the problem and synthesized a solution.

Shipping Gas

Your company has purchased a gas field in the coastal waters off Louisiana in order to have a supply of methane gas for your chemical plant on the western coast of Florida. Use Bloom's taxonomy to unravel a plan to get the gas from the field to the chemical plant.

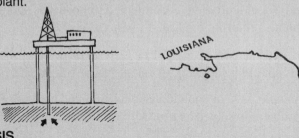

SYNTHESIS

CONSTRUCTING
THE SOLUTION

The first problem statement is: *"Find a way to transport the gas from Louisiana to Florida."* We begin this task by generating ways to accomplish the transportation. The techniques in Chapter 4 will greatly aid us in addressing this first task. Some of the ways this might be accomplished are to build a pipeline from the field to the plant, to ship by rail, or to ship by sea. A K.T. Decision Analysis shows that building a pipeline is too expensive and that the plant is not anywhere near a rail line. Consequently, the transport of the gas will take place by liquefying the gas and then shipping it to the chemical plant by boat. The next statement is: *"Design a system that will liquefy 2000 lbs of methane per hour for shipment by boat from Louisiana to Florida."*

EVALUATE

Before proceeding further to carry out a detailed design and sizing of the various pieces of equipment, we need to pause and evaluate the overall scheme. That is, we need to stop and do a preliminary evaluation of the proposal to ship compressed methane to the Florida plant. In the next chapter, we discuss the various items to consider in the **evaluation** phase. For example, is this scheme reasonable? What does a Potential Problem Analysis reveal?

ANALYSIS

In the analysis step we break the problem into parts and then examine each part. In order to liquify the methane, it must be compressed and cooled. The liquid methane will then be pumped into the ship. After shipping, the methane will be off-loaded at the plant in Florida.

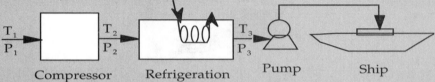

In this example, an analysis reveals the parts are a compressor, a heat exchanger, a pump, and a ship. Having recognized that we need a compressor to liquefy the gas, we collect information to learn what pressure and temperature are necessary to liquefy and transport the methane. Next we determine which type of compressor we should use. Should it be a centrifugal or a reciprocating pump?

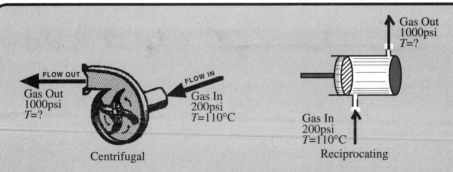

Centrifugal   Reciprocating

CONSTRUCTING
THE SOLUTION
(cont'd)

The gas enters the compressor at 200 psi and 110°C, and is to be compressed to 1000 psi. In order to design the heat exchanger, we must know the temperature of the gas as it leaves the compressor and enters the heat exchanger. Other points to be addressed at the analysis stage are the number of pumps required, and whether interstage cooling is required between the pumps. (K.T. Decision Analysis was used and a reciprocating pump operating close to adiabatic conditions was chosen.)

APPLICATION

We are going to use a reciprocating pump compressor and we need to know the temperature exiting the pump before we can design the heat exchanger. In the application stage, we recall the laws that apply and what assumptions are reasonable.

For adiabatic operation we recall (or look up) the pressure-volume relationship for a gas compressed adiabatically:

$$PV^{\gamma} = \text{constant}$$

We also recall the ideal gas law: $PV^{\gamma} = RT$

where

P = pressure, kPa, V = specific volume, m³/mol, T = temperature, K

γ = ratio of specific heats = C_p/C_v, R = Ideal Gas Constant, (m³•kPa)/(mol•K)

Note: We could also have considered departures from ideal gas law behavior.

CARRY
THROUGH
(UNRAVELING
TO COMPLETE
THE SOLUTION)

COMPREHENSION

Here we manipulate the equations in order to predict the exit temperature from the compressor.

$$T_2 = T_1 \left[\frac{P_2}{P_1} \right]^{\left(1 - \frac{1}{\gamma}\right)}$$

$$T_2 = 383 \text{ K} \left[\frac{1000}{200} \right]^{\left(1 - \frac{1}{1.2}\right)} = (383 \text{ K})(1.31) = 501 \text{ K} = 228°\text{C}$$

EVALUATION

Are the numbers reasonable? Use the checklist in Chapter 7 (p. 151) to evaluate the solution. A check of related problems in thermodynamics texts shows that this is indeed a reasonable number.

6.4 FOLLOW UP

Flexibility is an essential trait for problem solvers to have in order to deal with the inevitable changes that occur during projects. Finally, in the *Follow Up Phase*, we monitor not only our progress with respect to time deadlines but also with respect to meeting solution goals that do indeed solve the problem. In this phase we periodically check the progress of the *Carry Through Phase* to make sure it is

☞ following the solution plan
  • meeting solution goals
  • fulfilling solution criteria
☞ proceeding on schedule
☞ within budget
☞ of acceptable quality
☞ still relevant to solving the original problem

> *Inspect what you Expect.*

It is important to check these points to make sure the solution is "on track" and satisfying all the necessary goals. Be sure to check periodically that the problem is still correctly defined during the implementation phase. Sometimes a change in conditions can occur during implementation that will invalidate the solution.

6.5 PROBLEM STATEMENTS THAT CHANGE WITH TIME

Sometimes it may feel as if you are shooting at a moving target, as the desired goals change over the course of the project. A change in the problem statement could be the result of changing market conditions, the introduction of a competing product or services, reduced financing, or other factors. If during the *Carry Through Phase* some part, or perhaps all of the project cannot be accomplished, the problem statement must be modified. This type of information is only learned *after* we begin the solution. For example, during the course of your product development, your competitor markets a more advanced model than you were designing for nearly the same price. Consequently, you are now faced with several alternatives, which include cutting your price to significantly undercut that of the new product, or improving your design to surpass that of your competitor.

When the Goal Keeps Changing?

In the 1870s, a military fort was built in the west near a small village on the northern plains. As the first winter approached, the captain of the garrison sent his men to the forest to obtain firewood. The *initial goal* was to chop down enough trees to stack eight cords of wood. The captain then asked the corporal to ride over to the village to ask an old settler how cold the winter would be in order to determine if he had cut a sufficient amount of wood to last the winter. When the corporal arrived at the village and asked the old settler, the old settler put his hand to his forehead, looked towards the fort and said, "**Cold winter!**" The corporal reported this to the captain and as a result, eight more cords of wood were cut and stacked.

The captain then asked the corporal to check once again with the old settler as to how cold the winter was going to be. The old settler looked to the sky in the direction of the fort and this time said, "**Cold, cold winter!**" When the corporal reported to the captain, the captain ordered eight more cords of wood cut and then asked the corporal to check once again with the old settler. The old settler looked towards the fort and said, "**Cold, cold, cold winter!**"

The corporal said, "Wait a minute. The first time you said 'Cold winter,' the second time, 'Cold, cold, winter,' and now you say, '**Cold, cold, cold winter.**' Why do you keep changing your mind saying the winter is going to be colder and colder?" "Because," said the old settler, "the people at the fort keep stacking more and more wood for the winter!"

If the goals keep changing, keep two things in mind: 1) Where did the goals come from and why? 2) Are the goals still appropriate to the problem as originally defined? Be flexible and make adjustments as necessary.

6.6 EXPERIMENTAL PROJECTS

To solve a problem, you sometimes need more information than you have available. In fact, the specific information you need may not even exist (or you can't find it in a timely fashion). Therefore, you may be required to initiate an experimental program in order to generate the necessary data or information. How do you do this in an efficient manner? The following figure maps out an efficient path for an experimental program.

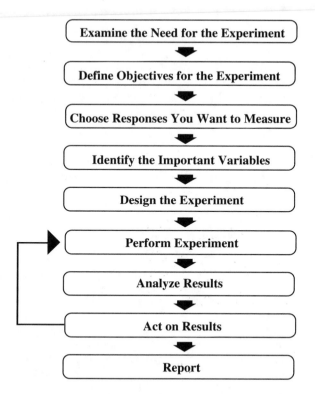

Figure 6-3. Flow Chart for Experimental Projects

6.6A Do You Really Need the Experiments?

When you are preparing to initiate an experimental program, be sure to question yourself and others to help guide your progress. The following questions will help you dig deeper into your project.

- Why perform the experiments?
- Can the information you are seeking be found elsewhere (such as literature journals, books, company reports, etc.)?
- Can you do some calculations instead?
- Have sufficient time and money been budgeted for the program?
- Are you restricted to specific materials or equipment?
- Will the safety of the investigators be endangered to such a degree that the program should not be carried out?

These and other appropriate questions must be answered prior to beginning the experimental program so that the need for the experiments is clearly established.

6.6B Define the Objectives of the Experiment

Prepare a list of all the things you want to accomplish. Next try to prioritize your list, keeping in mind the following:

- What questions regarding your problem would you most like to answer?
- Are you sure you are not losing sight of the overall objectives and other possible alternative solutions ("Can't see the forest for the trees" syndrome)?
- How comprehensive does the program need to be? Are you looking at an exhaustive study or a cursory examination of a narrow set of conditions?

Specific answers to these questions will guide the rest of the project.

6.6C Choose the Responses You Want to Measure

There are generally two different types of variables that are considered in an experimental program. The *independent* variables make things happen. Changes in the independent variables cause the system to respond. The *responses* are the *dependent* variables. The film speed, flash, and focus are all independent variables related to a camera. Changing any one of these will change the system response, i.e., the quality of the picture (the dependent variable). As the experimental program is designed, the important dependent variables to be measured must be identified.

- What are the controlled or independent variables?
- What are the dependent variables?
- Are instruments or techniques available to make the measurements available?
- Do they need to be calibrated? If so, have they been?
- Will the accuracy and precision of the expected results be sufficient to distinguish between different theories or possible outcomes?

6.6D Identify the Important Variables

In any experimental program there will always be many, many quantities you can measure. However, you must decide which independent variables have the greatest influence on the dependent variable.

- What are the *really* important measurements to make?
- What are the ranges or levels of these variables to be examined?
- Instead of changing each independent variable separately, can dimensionless ratios or groups be formed and varied so as to produce the same end results with fewer measurements? (See Appendix A2.5.)

A Night at the Movies

One cannot go to a movie theater without seeing popcorn being sold. James Wilson is an entrepreneur who owns a moderate-size company that markets and packages a gourmet popcorn. He believes his popcorn is superior to that being sold at the movies and hopes that he can sell it to the major chains at a cost competitive with their current brand. While James maintains his popcorn tastes better (and will even vary with the type of movie–horror vs. musical),[†] his real selling point is that he gets a far greater percentage of his kernels to pop. He decides to carry out a series of experiments that he can use to present to the management of the major cinema chains (e.g., Showcase) to convince them to buy his gourmet popcorn.

A. **Establish a Need for the Experiment**. James believes he has a better product, yet no data are available that compare the popping efficiency of his gourmet popcorn with the brand currently sold at theaters.
B. **Define the Objectives of the Experiment**. The objective is to show that James's gourmet popcorn has a greater popping efficiency under a variety of popping conditions.
C. **Choose the Responses to Be Measured**. Measure the number (or fraction) of kernels that pop.
D. **Identify the Important Variables**. The major variables will be (1) age of the popcorn and (2) the mediums used to pop the corn (oil or air).

We will continue this example after further discussion of experimental design.

[†]The hypothesis has indeed been put forth that the type of movie being viewed has an effect on the taste of the popcorn. Perhaps you could prepare an experimental plan to test this.

6.6E Design the Experiment

To obtain the maximum benefit from a series of experiments, they must be properly designed. How can the experimental program be designed to achieve the experimental objectives in the simplest manner with the minimum number of measurements and the least expense? A *successfully designed experiment* is a series of *organized* trials which enable you to obtain the most experimental information with the least amount of effort. Three important questions to consider when designing experiments are

- What are the types of errors to avoid?
- What is the minimum number of experiments that must be performed?
- When should we consider repeating experiments?

Types of Errors

There are two types of errors that should be avoided in experimental design. A Type I Error is one in which you declare that a variable has an effect on the experimental outcome, when in fact it really doesn't. A Type II Error occurs when we fail to discover a **real** effect. A Type II Error results in lost information; a variable gets incorrectly classified as insignificant to the process or ignored and as a result, no further examination of it takes place. Type II Errors can be avoided by researching fundamental principles related to the experiments, gathering sufficient information, and planning thoughtfully.

James's Type I and Type II Errors

As an example of a *Type I Error*, consider the case of a series of experiments designed to determine the "popping efficiency" (defined as the percent of kernels popped) of a new brand of popcorn developed by James Wilson. James had growers producing his popcorn in different parts of the country, and he was interested in the "popping efficiency" of the corn from different growers. James tested the corn samples provided by the various growers of his new strain of corn and he concluded that two of the farmers from the state of Iowa produced corn with a significantly higher popping efficiency than growers from other locations. This result led James to the conclusion that the quality of the popcorn is very sensitive to the growing locations, and he should concentrate on producers in Iowa to supply his corn. Further investigation proved this to be a *Type I Error*. That is, the key variable that James overlooked was that the Iowa farmers who supplied the superior corn took great care to preserve the freshness and hence moisture content of the kernels in the final product. When the farmers from other locations took the same precautions to prevent excess drying as a result of long storage times, their corn popped equally well. Thus, James's declaration that growing location is a significant variable was a *Type I Error*. The usual penalty for such an error is a loss of credibility. A *Type II Error* would exist if James failed to realize that the age and the corresponding moisture content of the unpopped kernels were significant process variables.

The Minimum Number of Experiments
(or. . . "Getting the Most Bang for Your Buck")

The minimum number of experiments that must be performed is related to the number of important independent variables that can affect the experiment and to how precisely we can measure the results of the experiment. In designing the experiments, we will first choose two levels (i.e., settings) for each independent variable. Because these levels are usually at the extremes of the variable range, we refer to these settings as high and low (e.g., on/off, red/green, 100 psi/14.7 psi, 100°C/ 0°C, etc.). For example, consider an experimental program where the dependent variable is a function of three independent variables (A, B, and C), each of which can take on two possible values or levels.

Independent Variable Names	Possible Levels
A	High Low
B	High Low
C	High Low

An automatic camera (one that automatically selects the shutter speed and aperture operating for given conditions) can have a number of variables associated with it, for example, film speed, flash, and focus. Each of these independent variables will affect the dependent variable, the picture quality. The levels of these variables are

Variable	Symbol	High (+)	Low(−)
Film Speed	(A)	400 ASA	100 ASA
Flash	(B)	ON	OFF
Focus	(C)	IN	OUT

If all possible variable combinations were to be tested, the number of experiments is equal to the number of levels, N, raised to the power of the number of independent variables, n. For the camera example, the number of experiments necessary to test all combinations of independent variables is equal to $N^n = 2^3 = 8$ experiments. These are detailed below ((+) indicates a high level, while (−) indicates a low level of a particular variable).

Experiment No.	A	B	C
1	−	−	−
2	+	−	−
3	−	+	−
4	−	−	+
5	+	+	+
6	−	+	+
7	+	−	+
8	+	+	−

If there is no interaction among the variables (which may not be known beforehand), experiments 1–4 will yield all the necessary information. By no interaction, it is meant that each of the variables affects the outcome of the experiment independently and there is no synergistic effect of a combined interaction. When there is no interaction, the effect of variable A changing from a high value to a low value is always the same, regardless whether or not the values of B and C are high or low. For example, there *is no interaction* between the focus and flash because if the camera is out of focus, it will be out of focus for both cases of the flash (ON/OFF) and it will also be out of focus no matter what film speed is used. On the other hand, there *is* interaction between the flash and the film speed because a flash might be necessary (ON+) at low film speeds but not at high film speeds (OFF −).

Experiments 1–4 explore the effect of raising each variable, in turn, from its low level to its high level. In this type of situation, the minimum number of experiments that must be run is the number of independent variables plus one (3 + 1 = 4). Thus, *if there is no interaction*, we need only examine each variable individually at the high level, and we can predict the results of the other experiments (5 through 8) by combinations of the appropriate responses.

Experiment No.	A	B	C	Comments
1	–	–	–	Base Case
2	+	–	–	Reveals effect of high A
3	–	+	–	Reveals effect of high B
4	–	–	+	Reveals effect of high C

A full factorial design (all eight experiments in this case) is useful for developing a model to predict the outcome of experiments whose independent variables can change continuously (that is, they can assume a continuous range of values and not just two discrete values). Two levels (at least) of each of the variables are examined and the results can be interpreted in the form of a model to predict the outcome of future experiments. Deming[2] discusses this method of statistically designing experiments.

A Night at the Movies–The Sequel

To illustrate the method for determining the minimum number of experiments, let's return to James Wilson, the popcorn entrepreneur, and help him design a series of experiments to measure the popping efficiency of his new popcorn.

James would like to compare his new gourmet popcorn with a popular brand sold nationally. He has identified two important process variables he would like to consider: degree of freshness and method of popping (hot oil versus hot air). He plans to use accelerated aging by heating the fresh kernels in an oven at 75°C for a fixed period of time to simulate the aging process.

Let's consider a series of experiments that James might perform, using two levels of each of the three independent variables. We will indicate the high level of a variable with a plus sign (+) and a low level with a minus sign (–).

Independent Variable	Low Level (–)	High Level (+)
Popcorn Type	Theatre's (–)	James's Brand (+)
Freshness	Aged (–)	Fresh (+)
Popping Method	Hot Oil (–)	Hot Air (+)

- continued -

A Night at the Movies– continued

Note that the classification by level is arbitrary; it serves only to indicate that each independent variable of interest can have two different levels (settings). Remember, what James is interested in is the effect of these three independent variables on the "popping efficiency" (kernels popped per hundred kernels). If we were to check all possible combinations of these variables, we would need to perform eight different experiments:

Experiment No.	Popcorn Type	Freshness	Popping Method
1	Theater's	Aged	Oil
2	James's	Aged	Oil
3	Theater's	Fresh	Oil
4	Theater's	Aged	Air
5	James's	Fresh	Air
6	Theater's	Fresh	Air
7	James's	Aged	Air
8	James's	Fresh	Oil

This covers all possible combinations. It is easier to "see" the pattern if we repeat the table using the (+) and (–) indications of the level.

Experiment No.	Popcorn Type	Freshness	Popping Method
1	–	–	–
2	+	–	–
3	–	+	–
4	–	–	+
5	+	+	+
6	–	+	+
7	+	–	+
8	+	+	–

The pattern is now clear: The first four experiments examine the base case (low, low, low) and the effect of raising each variable in turn from its low level to its high level. The second four experiments examine the base case (high, high, high) and the effect of lowering each variable in turn from its high level to its low level.

If there is no interaction among the variables, then we can "get by" with only doing four experiments (either 1–4 or 5–8). For example, if aging has the same effect on all types of kernels, we need only examine the effect of aging on either James's or gourmet popcorn to determine its effect on all kernels. Hence, we could gather the necessary aging information from experiments 1 and 4, while runs 5 and 8 would not be necessary. If, however, there was some interaction among the variables, then all eight runs would be necessary. For example, let's suppose one of the kernel types has a thicker shell that is more resistant to moisture loss during aging. The effect of aging would then be different for the different kernels and we would require the additional runs to explore the interactions.

6.6F Performing the Experiment: How Many Times?

If there is some error associated with measuring the outcome of an experiment, we must consider repeating some of the trials to be sure we have accurate information. But how much data is enough? The answer to this question depends on how precise (reproducible) the experiments are and on how small a change in the outcome or result of an experiment we wish to detect. Obviously, the less precise the measurements (i.e., the more error that is present) and the smaller the change we are interested in, the more data we must collect and average to be confident in our result. In the popcorn example, the kernels are inherently nonuniform, and it may be easy to draw erroneous conclusions from a limited number of results. Averaging several runs under the same conditions is the best way to deal with such a situation to ensure reliable results. The required number of times that each run should be repeated prior to averaging can easily be calculated using a statistical procedure discussed by Hendrix.[3]

6.6G Analyze the Results

How good are the measurements? What modifications, if any, of the existing equipment are necessary to improve the accuracy or precision of the measurements or to better achieve the overall experimental objectives? In the popcorn experiments, we might find that the aging results are inconclusive and that we need to control the humidity more tightly in the oven during the accelerated aging process, and additional runs would be necessary.

Is there software available to perform least-squares analysis (see Appendix 2), set confidence limits, or other statistical analyses? Is there any mathematical model or theory available that suggests how the data might be plotted or correlated? What generalizations can be made from the data? Should other experiments be run to extend the data into different regions? Has an error analysis been performed, sources of error listed, and discussed in relation to how they affect the final result (i.e., by what magnitude and in what direction?). Finally, **have all experimental objectives been satisfied**?

Popcorn Engineering (P.E.)
Analyze the Results

The results of James Wilson's popcorn efficiency testing were as follows: Note that because of the variable nature of the kernels, James repeated each of the experiments several times.

Experiment No.	Popcorn Type	Freshness	Popping Method	% Unpopped					
	Theater's Brand (−)	Aged (−)	Oil (−)	Runs					
	James's Brand (+)	Fresh (+)	Air (+)	1	2	3	4	5	Avg
1	−	−	−	22	18	19	26	25	22 ± 4.4
2	+	−	−	16	15	15	14	15	15 ± 0.9
3	−	+	−	15	17	16	13	14	15 ± 1.9
4	−	−	+	19	18	18	21	17	18.6 ± 1.8
5	+	+	+	4	7	2	5	3	4.2 ± 2.4
6	−	+	+	12	14	15	13	13	13.4 ± 1.4
7	+	−	+	10	9	13	11	10	10.6 ± 1.8
8	+	+	−	3	5	5	7	4	4.8 ± 1.8

The mean percent unpopped kernels are shown in the right-hand column of the table for each set of experimental conditions. The variations about the mean (± value shown) were calculated using the student t-test (see Appendix A2.4, p. 192). For illustrative purposes, the means are plotted on bar charts shown below for the two different types of popcorn.

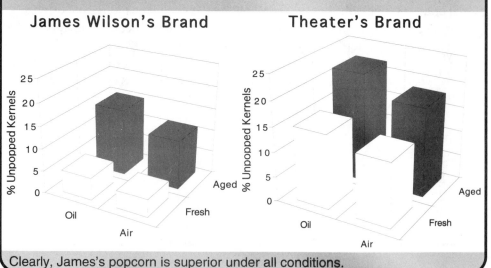

Clearly, James's popcorn is superior under all conditions.

Even though James's brand is clearly superior to the Theater's brand under all conditions, there appears to be a significant effect of aging that he is interested in exploring before taking his information to the theater chain. He decides to design a series of experiments to study aging of the kernels. The accelerated aging experiments will be performed in a constant temperature, constant humidity environment (oven) for varying lengths of time. Additionally, because of the inherently variable nature of the popcorn kernels, several samples will be taken for each set of conditions. The results of James's experiments are summarized below. The error bars (the ± values) were calculated using the *t*-test (see Appendix A2.4, pp. 192–194).

Accelerated Aging Tests on James Wilson's Popcorn

Percent Unpopped
(James Wilson Brand–Air Popped)

Days Aged	Run 1	Run 2	Run 3	Run 4	Run 5	Average	(±)
0	4	5	5	4	3	4.2	1.04
1	7	6	8	7	7	7	0.87
2	9	9	8	8	7	8.2	1.04
5	8	9	9	10	11	9.4	1.41
10	11	13	13	12	14	12.6	1.41
15	16	17	16	15	17	16.2	1.04
20	19	21	21	20	19	20	1.24
25	24	23	23	22	24	23.2	1.04
30	25	25	27	23	24	24.8	1.85

Percent Unpopped
(James Wilson Brand–Oil Popped)

Days Aged	Run 1	Run 2	Run 3	Run 4	Run 5	Average	(±)
0	5	5	5	5	4	4.8	0.55
1	9	8	8	7	8	8	0.87
2	10	11	10	12	9	10.4	1.41
5	13	12	13	12	11	12.2	1.04
10	16	15	15	16	17	15.8	1.04
15	18	19	19	21	17	18.8	1.85
20	21	23	23	22	24	22.6	1.41
25	24	25	26	26	25	25.2	1.04
30	27	26	26	27	28	26.8	1.04

The average values along with the error bars are shown in the following graph.

Results of Aging Tests on James Wilson's Popcorn

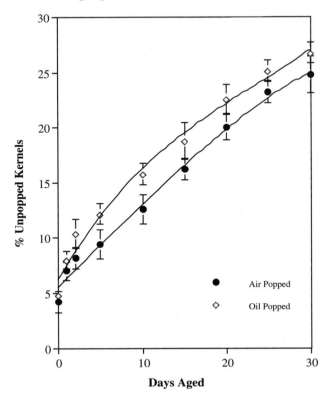

The graph indicates that there indeed is an effect of aging on the kernels. James postulates that this phenomenon is related to the kernels coming to equilibrium with the temperature/humidity conditions in the oven. There also appear to be similar effects of aging on oil and air popping . The error bars indicate that air popping seems to provide a more efficient means of popping the kernels, but the effect is not very great.

6.6H Report

Communicate the results of your work with other members of your team. This is usually done by means of a technical report. Guides for writing such a report can be found in many books. One good source is *Designing Technical Reports* by J.C. Mathes and D.W. Stephenson (Indianapolis, Bobb-Merrill, 1976), which has particularly useful examples. Typically a report will include the following sections:

1. *Abstract*–This one-page summary of the report is usually written last. It defines the problem, tells how you approached the problem, and the important results that were found.

2. *Introduction* –The introduction section defines the problem, tells why it is an important problem worthy of being studied, gives background information, describes the fundamental issues, and discusses and analyzes how they relate to published work in the area.

3. *Materials and Methods* –This section describes the equipment used to carry out the experiments, as well as instruments used to analyze the data. The purity of the raw materials is specified, as are the brand names of each piece of equipment. The accuracy of each measurement taken is discussed. The step-by-step procedure as to how a typical run is carried out is presented, and all sources of error are discussed. (If you developed a new model or theory, then a *Theory Section* would come after section 3. The theory section would develop the governing equations that mathematically describe your phenomena and justify all assumptions in the development.)

If you don't effectively communicate your results, you may as well have not performed the experiment.

4. *Results* –This section tells what you found. Make sure figures and tables all have titles and the units of each variable are displayed. Discuss all sources of error and describe how they would affect your results. Put an error bar on your data where appropriate.

5. *Discussion of Results* –This section tells why the results look the way they do. Discuss whether they are consistent with theory, either one you developed or that of others. You should describe where theory and experiment are in good agreement as well as those conditions where the theory would not apply.

6. *Conclusion* –The conclusion section lists all important information you learned from this work in numerical order, e.g.,
   1. The reaction is insignificant below 0°C.
   2. The results can be described by the Buckley-Leverette Theory.

7. *References* –List all resource material you referred to in this work in the proper bibliographical format.

In addition to the above sections of the technical report, many companies require an executive summary which is usually an expanded version of the abstract that includes the conclusion and recommendations. However, at least of equal importance to the sections that are in the report are the *style* and *clarity* with which the report is written. A top ten list of things to look for in an effective report is shown on the next page.

TOP TEN LIST FOR EFFECTIVE WRITTEN REPORTS

1. No Disagreements with Strunk and White[*] (i.e., perfect grammar)
2. Logically Organized (Introduction, . . . , Conclusions)
3. Logical Flow of Ideas within Each Section
4. Concisely Written
5. Interestingly Written, Using Wide Variety of Words
6. Ideas Supported by Examples, Data, Evidence
7. Appropriate Use and Placement, Figures and Tables
8. Passive Voice
9. Clear Purpose
10. Put in a Clear Plastic Binder (per Calvin and Hobbes)

In addition to written reports, you will also be expected to give oral reports throughout your career and you should refer to some of the references at the end of the chapter to prepare your presentation. The top ten list below identifies some of the key points to consider.

Tell em what your gonna tell em, tell em, and tell em what you told em.

TOP TEN LIST FOR EFFECTIVE PRESENTATIONS

1. Well Organized (Introduction, Body, Closure)
2. Logical Flow of Ideas
3. Ideas Presented Concisely
4. Ideas Supported by Examples, Data, etc.
5. Clear Explanations
6. Good Visual Aids
7. Speak Clearly and at a Reasonable Speed
8. Well Prepared and Practiced
9. Dress Appropriately
10. Conclusions Supported by Evidence
11. Match Presentation to Audience
12. Confident Appearance
13. Good Dictation and Grammar–Avoiding Slang
14. Disregard Three of the Above to Make It a Top Ten List

---

[*]Strunk, W. and E.B. White, *The Elements of Style*, Macmillan Publishing Company, Inc., New York, 1979.

CLOSURE

The first step in the Implementation Process is planning. Here we use *Gantt* and *Deployment* charts along with *Critical Path Planning* and *Resource Allocation* to guide us as we prepare our solution. In some cases, the objectives, goals, or even the problem statement may change while we are in the process of carrying through the solution. In this phase, all steps of the heuristic are reviewed.

SUMMARY

- *Planning*
  - Sketch the pathway through to the solution.
  - Prepare Gantt and Deployment charts.
  - Identify the critical tasks and prepare a Critical Path Diagram.
  - Prepare a budget for your project.

- *Carry Through*
  - Monitor the progress of the critical tasks closely.
  - Use the Carry Through check list.
  - Be alert for possible changes in the original problem statement and chosen solution.
  - Use Bloom's Taxonomy to unravel the solution.

- *Follow Up*
  - Check to see if the solution meets the specified objectives and criteria.

- *Experimental Projects*
  - Formulate your experimental plan carefully to maximize your efforts.
  - Be aware of the types of errors that can occur.
  - Estimate the minimum number of experiments to carry out.

REFERENCES

1. M.S. Peters, and K.D. Timmerhaus, *Plant Design and Economics for Chemical Engineers*, 4th ed., McGraw-Hill Publishing Co., New York, 1991.
2. Deming, S.N., "Quality by Design, Part 5," *Chemtech*, p. 118, February 1990.
3. Hendrix, Charles D., "What Every Technologist Should Know about Experimental Design," *Chemtech*, p. 167, March 1979.

FURTHER READING

Starfield, Anthony M., Karl A. Smith, and Andrew L. Bleloch, *How to Model It. Problem-solving for the Computer Age*, McGraw-Hill Publishing Co., New York, 1990. Chapter 6 of this book contains a nice description and additional examples of critical path planning.

Massinall, John L., "The Joys of Excellence," *Chemtech*, 20, p. 393, July 1990. This article has some timely tips for managing a project and keeping work on schedule.

Hendrix, Charles D., "What Every Technologist Should Know About Experimental Design," *Chemtech*, p. 167, March 1979. Some useful background material on designing experiments to gain the maximum amount of information with the least amount of effort.

McCluskey, R. J., and S.L. Harris, "The Coffee Pot Experiment–A Better Cup of Coffee via Factorial Design," *Chemical Eng. Educ.*, p.151, Summer 1989. More interesting background on designing experiments.

Blanchard, Kenneth, and Robert Lorber, *Putting the One Minute Manger to Work–How to Turn the 3 Secrets into Skills*, Berkeley Books, Berkeley, CA, 1984. Increase your productivity using three easy to follow techniques.

Murphy, Thomas D., "Design on Analysis of Industrial Experiments," *Chemical Engineering*, 84, p. 168, June 6, 1977. Excellent overview of factorial design of experiments.

EXERCISES

Consider the following situations:
   A. Planning a surprise birthday party
   B. Planning a wedding
   C. Planning a camping trip to Colorado
   D. Getting elected mayor of your city
   E. Publishing your autobiography
   F. Becoming a Supreme Court justice

For Problems 1 through 6, choose situation A, B, C, D, E, or F from the above list.

1. Prepare a Gantt Chart on _____above.

2. Prepare a Deployment Chart on_____above.

3. Prepare a Critical Path Planning Chart on _____above.

4. Use Bloom's Taxonomy to outline how you would carry through on _____ above.

5. Prepare a budget on _____ above.

6. Prepare a Gantt Chart, Deployment Chart, a Critical Path Planning Chart, and a budget on _____above.

7. Prepare a write-up (or presentation) to management to
   a. Attend a professional meeting or short course in Europe.
   b. Market a new widget.

8. You are going to prepare a three-course dinner for your gourmet dinner group for a party of eight.

Course	Item	Preparation time	Eating Time
Appetizer	Bacon-wrapped water chestnuts	Cook in oven 10 minutes	
Soup	Onion Soup	Cook 30 minutes on stove	15 minutes
	Bread Sticks	Warm 10 minutes in oven	
Entrée	Pot Roast	Cook 2 hours in oven	40 minutes
	Mashed Potatoes	Cook ready mix 10 minutes	
	Fresh Mixed Vegetables	Boil 20 minutes	
	Gravy	Cook juice from roast 10 minutes on stove	
Dessert	Apple pie	Cook 35 minutes in oven	15 minutes
	Ice Cream	Let stand 5 minutes before scooping	

Prepare a
      Gantt chart
      Stove Deployment Chart (i.e., oven top)
      Critical Path Diagram

9. Prepare an experimental plan to modify the popcorn example to include popping the corn in at least three different oils (peanut oil, corn oil, olive oil).

10. Design experiments to study the following problems given the important variables listed:

a) Coffee Making
    • grind (coarse vs. fine)
    • type of bean (standard vs. premium)
    • percolator vs. automatic drip

b) Paper Airplane Making
    • wing span
    • weight distribution
    • initial velocity
    • angle of takeoff

c) Plant Growth
    • hours of sunlight
    • amount/frequency of watering
    • fertilizer use

d) Paper Production (quality of product . . . strength, brightness)
    • whiteness (use of bleaching agents)
    • Percent recycled fibers used
    • types of fibers used (new and recycled)

e) Fementation of Sugar to Alcohol by Yeast
    • temperature
    • pH
    • nutrients
    • desired final alcohol content

7 EVALUATION

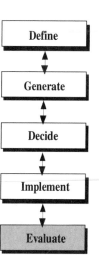

After we have implemented our solution, we need to perform a final evaluation of our solution. In this chapter we present guidelines for evaluating our solution to make sure it 1) completely solves the problem specified, 2) is ethical, and 3) is safe to people and the environment.

7.1 GENERAL GUIDELINES

Evaluation should be an *ongoing* process throughout the life of a project. As each phase of the project is completed, the goals and accomplishments of that phase should be examined to make sure they were satisfied before proceeding to the next phase. Evaluate future directions in light of the results of each phase to verify that the direction we are proceeding in is still the correct one. Look for any fallacies in logic that might have occurred, especially at key decision points during the project. Challenge the various assumptions that were made. Have all unstated assumptions been recognized? For example, was the engineer justified in assuming that the rate of cooling the chemical reaction products was the bottleneck for the entire process? To address these questions, have someone outside the group that developed the solution review the assumption and solution logic. After the decision has been reached, be sure to carry out and brainstorm a *K.T. Potential Problem Analysis* (PPA) before *carrying through* the solution. It is activities such as PPA that might have prevented many failures such as the Orecha River Dam project. During the evaluation process, qualitative and quantitative judgments about the extent to which the material and methods satisfy the external and internal criteria must be made. Ask a number of evaluation questions such as those given in the following checklist.

Evaluation Checklist

✓ Have you challenged the information and assumptions provided?
✓ Does the solution solve the *real* problem?
✓ Is the problem permanently solved, or is this a patchwork solution?
✓ Does the solution have impact?
✓ Have all the consequences of the solution (adverse as well as positive) been examined?
✓ Have you argued both sides–the positive *and* the negative?
✓ Has the solution accomplished all it could?
✓ Is the solution economically efficient and justifiable?
✓ Have the "customers" been surveyed to see if the solution meets all their needs?
✓ Does the solution cause other problems (e.g., environmental, safety)?
✓ Is the solution logical?
✓ Is the solution economically, environmentally, and politically responsible and safe?

The Snow Cruiser

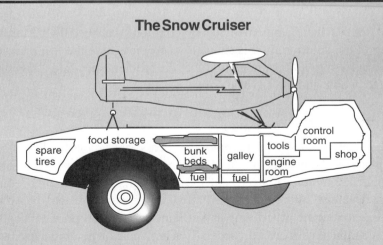

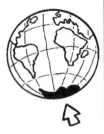

Following Admiral Byrd's second expedition to Antarctica in 1934, excitement over polar exploration was running high in the United States. Thomas Poulter, Admiral Byrd's second-in-command, had experienced the difficulties associated with polar expeditions firsthand and had some ideas about how to improve subzero working, living, and transportation problems. The embodiment of these ideas became the *Snow Cruiser*. In 1939, six months prior to the third U.S. expedition to the South Pole, Poulter, with help from over 70 private companies, set out to design and construct the Snow Cruiser. The vehicle was formidably designed to withstand some of the harshest environmental conditions on the face of the Earth. Some of the design features included:

- a range of 5000 miles
- room for a crew of five and enough supplies for one year
- an airplane carried on its roof for photographic missions
- outstanding terrain capabilities that included retractable tires for crossing small crevasses

The vehicle was constructed in Chicago, over a period of about three to four months (August to November). The design and construction period was quite rushed because of the expedition's scheduled "summer" arrival at the South Pole. The ships were to leave Boston for the Antarctic in November. The specifications of the completed vehicle were very impressive:

- 55 feet 8 inches long, 19 feet 10.5 inches wide, and 16 feet high
- gross weight (fully loaded) = 75,000 pounds

After construction, it was driven cross-country from Chicago to the port in Boston. The trip was quite eventful, and crowds gathered to watch the mammoth vehicle pass. Roads had to be closed to other traffic because of the great width of the Snow Cruiser. In Boston, it was loaded onto a ship, the North Star, to begin the expedition to Antarctica.

-continued-

The Snow Cruiser–continued

In January 1940, the North Star arrived at the Ross Ice Shelf in Antarctica. With great anticipation, the Snow Cruiser was driven off the ship by Thomas Poulter. Once on the snowy terrain, it became obvious that there was a problem. The Snow Cruiser could hardly move at all!! The big vehicle behaved much like a beached whale. The huge tires spun, unable to get traction. As they spun, heat was generated and the Snow Cruiser sank up to 3 feet in the snow. Something was seriously wrong with the design. Additionally, the vehicle was seriously underpowered. When it was able to gain some traction and move (which wasn't often) the engines overheated after only a few hundred yards, leaving the vehicle stranded again!

After several months of unsuccessfully trying to improve the vehicle's mobility, the Antarctic winter set in, and the expedition team gave up hope of using the Snow Cruiser for exploration. They covered it with timbers and snow and used it for shelter. The Snow Cruiser was an enormous flop.

What went wrong? Let's look at the failure of the Snow Cruiser (using 20/20 hindsight) and see how the some of the Evaluation Checklist questions on page 151 might have prevented this debacle.

Have you challenged the information and assumptions provided? What information is available about the Antarctic environment? How difficult is the terrain? If the vehicle will move on dry roads in good weather, what makes us think it will function on snow and ice? Most polar vehicles up to this time used caterpillar treads rather than tires. Why would our new tire design work? Why do other vehicles use treads? The answers to some of these obvious questions may help avoid a failed design.

Does the solution solve the real problem? What is the *real* problem in this case? Clearly the problem is at least twofold. One problem is to protect the workers and explorers from the harsh polar environment. The other, and just as important an aspect, is that the vehicle should have good mobility on the expected terrain so that exploration (the main goal of the expedition) is possible. The design was quite successful from the protection/living accommodations standpoint. The Snow Cruiser was nicer inside than many pre-World War II bungalows. The design only partially addressed the mobility problem. Elaborate design features were included to enable the Snow Cruiser to cross crevasses in the snow that it would certainly encounter, but it appears that insufficient consideration was given to ensuring "normal" mobility in polar ice and snow.

Surely some incorrect assumptions were made regarding the traction of the tires and the power necessary to move such a mammoth vehicle in these severe conditions. Challenging all the assumptions of the design and making sure that the real problem (and all facets of it) are solved are keys to determining a functional solution.

-continued-

> **The Snow Cruiser—continued**
>
> ***Is the problem permanently solved?*** If indeed the Snow Cruiser functions as designed, it would be a permanent solution to polar expedition problems.
>
> ***Does the solution have impact?*** In this case, yes. The Snow Cruiser could have revolutionized the way polar explorations were conducted.
>
> ***Have all consequences of the solution been examined?*** This question is difficult to answer, not knowing what went on at the time, but providing the vehicle operates as designed, it appears that many adverse consequences were anticipated and designed for. Provisions and fuel were available for long periods of time. It had a travel range of 5000 miles. Seemingly every contingency had been prepared for...except the fatal mobility flaw.
>
> ***Has a list of solution criteria been surveyed and your solution rated with respect to each criterion?*** This question might have brought to light some of the design problems, providing the problem was correctly defined in the first place.
>
> If used properly, and carefully, the Evaluation Checklist can point out some otherwise hidden flaws in the solution to a problem and catch costly "mistakes" before they are built.

In addition to the Evaluation Checklist, the McMaster Five-Point Strategy gives the following checklist for examining proposed solutions:

LIST

- Check that the solution is blunder-free.
- Check the reasonableness of results.
- Check that criteria and constraints are satisfied.
- Check the procedure and logic of your arguments.

You need to confirm **all** findings. Check to see if there is a piece of the puzzle (i.e., the solution) that doesn't fit and consequently may require the entire solution to be redone.

> **The Last Penny**
>
> Two close friends were seniors in college; one was a business major, the other an engineer. The business major was working on her senior accounting project, which was due in a few days, and was really frustrated because her accounting sheet for several hundred thousand dollars didn't balance by literally two or three dollars. The engineer pressed her as to why should she worry if she was off only a few dollars on a project budget of this magnitude. Her response was that it could be the result of two major errors that may have compensated one another. Consequently, when the last part of the puzzle didn't fit (i.e., the last few cents) the entire solution (i.e., the accounting project) had to be examined.

7.2 ETHICAL CONSIDERATIONS

Part of the evaluation process includes reviewing all facets of the solution to ensure that an ethical solution is in place. Many times the ethical aspects of the situation may not be entirely clear and must be uncovered in much the same manner as defining the real problem or discovering what the fault is in a troubleshooting operation. How many times have you heard someone say when discussing a conflict or arbitration, "I had not thought about it in that light?" A set of guidelines that may help us sort out the issues and guide us to an ethical solution has been addressed by Blanchard and Peale in their book *The Power of Ethical Management*.[1] They give the following checklist of questions to consider.

> Solutions are not always black and white with regard to ethics, but shades of gray.

The "Ethics Check" Questions[1]

1. Is it legal? Will I be violating either civil law or company policy?
2. Is it balanced? Is it fair to all concerned in the short term as well as the long term? Does it promote win-win relationships?
3. How will it make me feel about myself? Will it make me proud? Would I feel good if my decision were published in the newspaper? Would I feel good if my family knew about it?

If the answer to the first question could be interpreted from any viewpoint or appearance as "**NO**, it is not legal," then there is no need to proceed to the second and third *Ethics Check List* questions. However, if the solution is indeed legal and does not violate company policy, then the second question raises the flag that a decision that greatly benefits one person or company will eventually come back to haunt that individual or company. Blanchard and Peale's last question is meant to activate our sense of fairness and make sure that our self-esteem is not eroded through an unethical decision. This ethics check list helps us address one of the knottiest problems in business: "How can we get acceptable bottom line results, stay competitive, and at the same time make sure we are being ethical?"

> *"There is no pillow as soft as a clear conscience."*
>
> John Wooden, UCLA Bruins

In addition to the Ethics Check, Blanchard and Peale discuss the **Five P's** that need to be considered in analyzing the solution: Purpose, Pride, Patience, Persistence, and Perspective. Perspective on the whole problem is the key or central point. The **Five P's** table gives a list of questions for us to answer for each **P** that will help us further evaluate our solution.

The ethical evaluation process can be facilitated by choosing someone to discuss each of these questions. This person (called an advisor) could play a passive role simply listening to your explanation or an aggressive role by questioning your every point. Even in the passive mode, the mere fact that you verbalize the application of the **Five P's** to your situation will improve the evaluation process.

The Five P's

Purpose: What is the objective for which you are striving? Are you comfortable with that as your purpose? Does your purpose hold up when you look at yourself in the mirror?

Pride: Can you take pride in the solution you have developed? Is there any false pride or self-doubt involved?

Patience: Have you taken the time to think through all the ramifications of your solution?

Persistence: Are you sticking to your guns and not being dissuaded by other demands? Have you given up too soon on finding a solution that is fair and balanced to all concerned?

Perspective: Have you taken the time to focus inside yourself to be sure everything fits with your ideals and beliefs? How does the solution fit into the "Big Picture?"

Perspective is the fifth **P**, the hub around which the other **P**'s rotate. Part of *Perspective* is the inner guidance that is awakened from the other **P**'s that helps us see things more clearly.

"The greatest battles of life are fought out daily in the silent chambers of the soul."
—David McKay

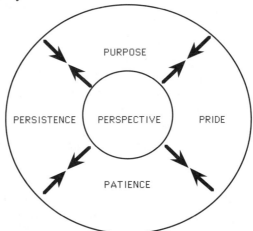

Figure 7-1. The Interrelationships Among the Five P's

As discussed earlier, one should carry out an evaluation at each of the key decision points and not wait until the very end of the project to learn that perhaps the solution path that one chose was unethical. We now consider the following examples.

The Christmas Gift

Henry is in a position to influence the selection of suppliers for the large volume of equipment that his firm purchases each year. At Christmas time, he usually receives small tokens from several salesmen, ranging from inexpensive ballpoint pens to a bottle of liquor. This year, however, one salesman sends an expensive briefcase stamped with Henry's initials. This gift is very much out of the ordinary.

Should Henry

1. Keep the case, on the grounds that his judgment will not be affected in any way?
2. Keep the case, since it would only cause embarrassment all around if the case were returned?
3. Return the case?
4. Other? (Please specify)

The Five P's

Purpose: Ask yourself what you would do if you were in Henry's shoes to remain unbiased in selecting the best supplier for a given job.

Pride: Would you feel pride in accepting the case or pride in returning the case?

Patience: Set aside a time to think about whether or not you should accept the case. Talk to someone whose judgment you trust.

Persistence: Have you pursued all avenues to resolve either keeping or returning the case?

Perspective: Even if you feel your judgment will not be affected by accepting the case, how will it appear to other colleagues? Are you setting a good example?

The following is the response to a reader survey carried out by *Chemical Engineering Magazine* (p. 132, Sept. 1980.). We note the majority of the people, 64.9%, thought the case should be returned. However, 27.7% of those under the age of 26 thought it would be OK to keep the case.

Option	U.S. Total	U.S. Total	U.S. (by age) % <26	U.S. (by age) % 26–50	U.S. (by age) % >50	Non-U.S., % Brit/Can	Non-U.S., % Others
1.	20.1	19.7	27.7	17.9	15.4	14.6	39.0
2.	3.4	3.3	4.9	2.8	3.1	4.9	4.8
3.	64.9	65.8	56.9	68.1	67.6	61.0	43.8
4.	9.8	9.5	9.0	9.8	12.1	18.3	11.0

Comments by the Responders:

1. "Keep the case and really not be affected by it, which means that next Christmas you're back on ball point pens."
2. "As a procurement agent on a limited scale, I have heard of colleagues receiving Porsches. The gifts only get bigger if you accept the first. Eventually, it will affect your judgment."
3. "I sometimes feel guilty accepting a gift knowing that the giver won't get anything out of it. But I won't be bribed."
4. "My price is very much higher than a briefcase. As a matter of fact it is so high that it would not be profitable to meet it."
5. "Henry risks being fired for a crummy $80 briefcase. That's the trouble, it's never two fully paid round-trip tickets to Hawaii, it's always cheap junk."

Rigging the Bidding

Steve is a designer for a large chemical company. There are two competitive pieces of equipment that are sold to do a job required by the process he is working on. Equipment from ABC Inc. is widely advertised and sold, but Steve has heard through the grapevine from his competitor's plant that ABC's equipment tends to break down unexpectedly and often. However, there is no way for him to document this information. XYZ Corp. makes equipment that will do the same job, but it is much more expensive. Steve knows from his own experience that this equipment is quite reliable. It is company policy to obtain competitive bids, and in such a situation, ABC would certainly win. Steve deliberately rigs the specifications by inserting unnecessary qualifications that only the XYZ equipment will meet.

Ethics Questions: Answer yes or no to each of the following options.
  A. Is Steve being ethical in using false specifications to circumvent company policy, even if he believes it is in the company's best interests?
  B. Would you do the same if you were in Steve's place?

The Five P's

Purpose: To choose the best piece of equipment for your company.

Pride: Will you feel pride in having reached the best decision and the way in which you reached it?

Patience: Have you taken the time to check with people in the company to seek their advice as to which way to go?

Persistence: Have you checked the grapevine to see if it is *rumor* or *fact* that the equipment is unreliable?

Perspective: Are you setting a precedent so that you might write future specifications for only one company?

Responses from the "Reader Survey" in *Chemical Engineering Magazine*, p. 132, Sept. 1980:

Question A	Total	U.S. Total	U. S. (by age) % <26	26–50	>50	Non-U.S., % Brit/Can	Others
1. Yes	36.8	36.8	40.8	17.9	32.8	32.9	41.8
2. No	55.7	55.7	53.5	57.0	54.9	56.7	54.1

Question B	Total	U.S. Total	U. S. (by age) % <26	26–50	>50	Non-U.S., % Brit/Can	Others
1. Yes	30.6	30.5	35.9	30.1	24.4	28.0	37.0
2. No	61.1	61.2	56.9	62.6	61.5	56.8	56.8

-continued-

The Five P's–continued

Comments by Responders:
1. "Can a professional judgment be made based on the grapevine?"
2. "Go get that grapevine story verified or busted. Get the facts."
3. "Anecdotal reports are a way to document suspicions. One must be careful that one isn't depending on a few disgruntled users; but, surely, it's reason to do more thorough research."
4. "To eliminate red tape within a large company and obtain the best equipment, one must often change the rules."
5. "Since it'd be me that was called at 2 a.m. when the ABC equipment broke, I'd probably force the choice of XYZ."
6. "I have not heard of any design engineer being made to decide based on price and specifications alone. Judgment is a highly sought quality in an engineer. Sometimes, gut feeling counts too."

7.3 SAFETY CONSIDERATIONS

One of the most important parts of the evaluation process is to make sure the proposed solution is safe to both humans as well as to the environment. Carrying out a Potential Problem Analysis or Hazard Operations Analysis (HAZOP) is helpful to make sure you have considered all aspects of the solution that might prevent it from being safe.

The Kerosene Heater–A Potential Problem Analysis

We are considering purchasing a portable kerosene heater to use in our family room this winter for additional heat on those long, cold nights. Let's consider some of the potential problems that might occur with the heater and construct a Potential Problem Analysis Table.

Potential Problem	Consequence	Possible Cause	Preventive Action	Contingent Action
Improper fuel used in heater	Explosion	Using gasoline to fill tank	Place warning sign on heater	Verify it's kerosene before lighting heater
Heater tips over	Fire	Children playing near heater	Place heater in protected location	Remove combustible materials from vicinity
Carbon monoxide buildup	Asphyxiation	Improper venting	Installed by qualified technician	Check vent prior to lighting
Carbon monoxide buildup	Asphyxiation	Chimney plugged	Check chimney each heating season	Regularly have chimney/flue cleaned

A method frequently used is to work backwards from the hazard or potential problem to determine all possible faults that could result in this failure. This procedure is repeated until all possible routes to the problem are discovered. A graphical representation of these pathways is called a Fault Tree Analysis. The diagram is constructed using some special symbols.

The **OR** symbol (or block) has two inputs, **A** and **B**. If *either* input exists (or is TRUE) then there is an output from the **OR** block. The **AND** block differs from the OR block, in that *both* inputs must exist (or be TRUE) to get an output. Thus, in the AND block, if either A or B is false, then there will be no output from the AND block. These symbols provide us with a convenient shorthand for constructing a Fault Tree Analysis Diagram. Let's return to the analysis of the kerosene heater for our family room.

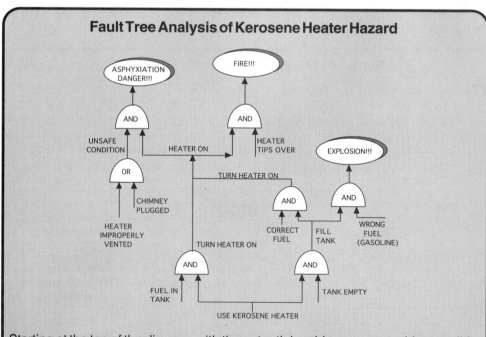

Fault Tree Analysis of Kerosene Heater Hazard

Starting at the top of the diagram with the potential problems, we consider possible scenarios that could result in those problems. For example, consider the explosion problem. One possible route to an explosion would be to light (turn on) the heater after it has been improperly fueled with gasoline. Note that both events must occur or no explosion will result, hence the use of the **AND** block. How could improper

-continued-

Fault Tree Analysis of Kerosene Heater Hazard–continued

fuel have gotten into the tank? It must have been empty *and* someone mistakenly filled it with gasoline. Let's consider another potential problem, asphyxiation. How could this happen? First, the heater must be on, *and* some fault must cause carbon monoxide to buildup in the house. What could cause that? Either a plugged chimney *or* an improperly vented heater could cause this situation (note the **OR** block connecting these two events). We continue to construct the diagram in this manner, trying to determine all potential problems and their causes. Once the potential problems are defined, we can devise preventive and contingent actions to guard against them.

Reaction in a Tank Reactor

A reaction is carried out in a reactor that operates at 500°F. The reaction is highly exothermic, meaning a significant amount of energy is released as the reaction proceeds. The feed to the reactor enters at 200°F and the energy released by the reaction just balances that necessary to heat the feed stream from 200°F to 500°F. Consequently, it is important to make sure there is not a decrease in the feed flow rate or an increase in feed temperatures, or else the temperature in the reactor will increase and the reactor could become unstable and explode.

The table below shows the *start* of our planning to prevent an accident in our reactor. First we consider the potential problem of no feed stream to the reactor. The consequences of interrupting the flow are that the reactor temperature will increase dramatically. One possible cause for no feed stream is the failure of the feed stream pump. We need to assign a probability that this could occur to help us prioritize the preventive actions we should take. For example, if the probability of occurrence is one in a billion (e.g., the pump is smashed by a falling object from the sky), then it is doubtful we will want to bother with any preventive action.

The preventive action for pump failure could be to have a parallel line from the feed tank to the reactor.

SOLUTION: Potential Problem Analysis

Potential Problem	Consequence	Possible Causes	Preventive Action	Contingent Action
No feed to reactor	Reactor temperature increases	Pump failure	Have auxiliary pump in parallel	Start diluent stream to reactor (shut down)
		Feed pipe ruptures	Inspect and test (replace) on a regular basis	Start diluent stream to reactor (shut down)
Feed temperature increases	Reactor temperature increases	Upstream temperature fluctuation	Place a surge tank upstream	Have auxiliary heat exchanger in feed line
				Add small shots of cold diluent to reactor

SUMMARY

- *Use the evaluation checklist to make sure you have*
  - Challenged information and assumptions.
  - Solved the *real* problem.
  - Examined all consequences of the solution.
  - A solution that is logical, ethical, safe, and environmentally responsible.

- *Use the Ethics Check List and the Five P's to make sure your decision/ solution is ethical*
  - Purpose
  - Pride
  - Patience
  - Persistence
  - Perspective

REFERENCES

1. Blanchard, K., and N.V. Peale, *The Power of Ethical Management*, Ballantine Books, Fawcett Crest, New York, 1988.

FURTHER READING

Dresner, M.R., "Risk Assessment: How Do We (Irrational Humans) Really Do It?," *Chemtech*, 21, p. 340, 1991. Provides an introduction to the subject.

Rosenzweig, Mark, and Charles Butcher, "Should You Use That Knowledge?," *Chemical Engineering Progress*, April 1992 and October 1992. Interesting ethical scenarios that further test your "ethical judgment."

Matley, Jay, and Richard Greene, "Ethics of Health, Safety, and Environment. 'What's Right'?," *Chemical Engineering*, p. 40, March 1987, and p. 119, September 28, 1987. More scenarios with a health/safety/environmental flavor to further challenge your ethical judgment.

EXERCISES

1. Nick is chief engineer in a phosphate fertilizer plant, which generates over a million tons per year of gypsum, a waste collected in a nearby pile. Over many years, the pile has grown into a 40-million-ton mountain. There being little room at the present site for any more waste, a new pile is planned.

    Current environmental regulations call for the elimination of acidic-water seepage and ground water contamination by phosphates and fluorides. Nick's design for the new pile, which has been approved, incorporates the latest technology and complies with EPA and state regulations.

    However, Nick knows that the old pile—although exempt from current regulation—presents a major public hazard. When it rains, acidic water seeps through the pile, carrying phosphates into the ground water.

In a confidential report to management, Nick recommends measures that will prevent this from happening. His company turns down his proposal, stating that, at present, no law or regulation demands such remedy. Use the *Evaluation Checklist*, the *Ethics Check List*, and the **Five P's** to help you analyze the situation. (From *Chemical Engineering Magazine*, p. 40, March 2, 1987)

2. The environmental and safety control group in a circuit-board etching and plating plant has just completed a program to improve the measurement of toxic releases to the atmosphere in response to stricter regulations recently issued by the state health and environmental commission.

   Small amounts of a toxic material are detected for the first time by means of a new instrument purchased and installed at the suggestion of Joan, the group leader. The detection method specified by the state does not reveal any trace of chemical.

   A search through books and magazines shows that this material is not dangerous in the low concentrations detected, although the state agency says it is, basing its claim on the extrapolation of published data. Use the *Evaluation Checklist*, the *Ethics Check List*, and the **Five P's** to help you adddress this situation. (From *Chemical Engineering Magazine*, p. 40, March 2, 1987)

3. Operators in a plant regularly dump over 100 drums a week of a dry, dusty enzyme (isomerase, used in converting dextrose to fructose) into a large tank, where it is rehydrated for several hours.

   The dust, although not immediately harmful, is suspected of causing long-term allergies and even lung problems. However, these symptoms show up in less than 10% of the people exposed.

   The plant safety code requires that operators wear masks, goggles, and gloves when dumping the enzyme. However, because the temperature in the working area is usually about 110°F, the operators ignore this requirement.

   This is the situation found by Phil, who has very recently taken over supervision of the department. Use the *Evaluation Checklist*, the *Ethics Check List*, and the **Five P's** to help you address this situation. (From *Chemical Engineering Magazine*, p. 40, March 2, 1987)

4. You are employed by a small company that is trying to build a plant to produce chemical X. As a result of your low overhead and other factors, you should be able to significantly underprice your competitor. You have sized all the pieces of equipment except the reactor(s). You cannot do this because you don't know the kinetic parameters nor do you have laboratory equipment to determine them. Time is of the essence, and your boss suggests that you take a photograph of your competitor's reactor, which is located outside their plant, and use that to size your reactor. He suggests you hire a plane for aerial photographs, or a truck similar to those used to repair telephone lines that could see over the fences. Your boss also suggests you could get an estimate of the production rate by monitoring the size and number of trucks shipping the product from the plant. How confident are you that your estimates of the sizes will be reasonably accurate? Do you feel it is ethical to estimate the reactor sizes and production rates in this manner? Make a list of reasons and arguments as to why your boss might feel it is ethical to request you to do this. If you don't feel this is ethical, make a list of reasons and arguments as to why it should not be done. Suggest alternative ways to obtain the desired information. You may wish to consult the book, *The Power of Ethical Management* by K. Blanchard and N.V. Peale (Fawcett Crest, New York, 1988) to identify and evaluate ethical issues. If you feel that the above situation is clearly ethical or unethical, revise the scenario so that it is in a gray area. For example, would it be ethical if the reactor were in full view from the street? What if your boss suggested that you get a tour of the plant with a Boy Scout troop, and try to take pictures and obtain other information (e.g., read gauges) while on

the tour at the competitor's annual "Engineering as a Profession Day" three weeks from now? Use the *Ethics Check List* to help take the grayness out of the situation and make it black or white.

5. You and your assigned partner, Brad, were supposed to complete a final report together for your laboratory course. You have just spent the entire weekend by yourself completing the group report which is due on Monday. You have been unsuccessful in reaching Brad and have left several messages on his answering machine. Brad finally shows up at the computer center at 8:00 PM on Sunday just as you are finishing the report, and tells you he decided to go home to visit his parents. Since this is a group report, you and Brad will receive the same grade for your efforts. You are disgusted with Brad for his lack of contribution to the report. How do you deal with the situation? Your lab partner? Your instructor? Make a list of suggestions to resolve or prevent the problem. What if your lab partner was your significant other and was out partying the entire weekend? Use the *Evaluation Checklist, Ethics Check List,* and the **Five P's** to help you answer these questions. (Developed by Bradley Foerster, University of Michigan, Class of 1992)

6. You are a project engineer at the Fine Chemical Company. You and Joe have been assigned to study the recently developed cobalt catalyst. You have just completed a six-month study of the cobalt catalyst and have been assembling a report. You have put in long hours this year, partly to make up for Joe's lack of effort and partly to receive the one promotion rumored to be available in your department. Your boss, Dr. Adams, has been anxiously awaiting your report. As you walk down the hall to the coffee machine, you overhear your partner discussing the cobalt catalyst in Dr. Adams' office. As they step into the hall, Dr. Adams slaps Joe on the back for his "dedicated and superior work on the cobalt catalyst" and promises to remember Joe during the end of the year evaluations. Joe has claimed your work as his own. What do you do? Propose possible approaches you could take with Joe and Dr. Adams to receive the credit you deserve. How would you deal with the situation if Joe's lack of effort were due to substance abuse? If the Fine Chemical Company were laying off engineers, would you be more concerned with receiving credit for the report or keeping your job? Use the *Evaluation Checklist, Ethics Check List,* and the **Five P's** to help you answer these questions. (Developed by Bradley Foerster, University of Michigan, 1992)

7. Jay's boss is an acknowledged expert in the field of data analysis. Jay is the leader of a group that has been charged with developing a new catalyst system, and the search has narrowed to two possibilities, Catalyst A and Catalyst B.

   The boss is certain that the best choice is A, but he directs that the tests be run on both, "just for the record." Owing to inexperienced help, the tests take longer than expected, and the results show that B is the preferred material. The engineers question the validity of the tests, but because of the project's timetable, there is no time to repeat the series. The boss directs Jay to work the math backwards and come up with phony data to substantiate the choice of Catalyst A, a choice that all the engineers in the group, including Jay, fully agree with. Jay writes the report.
   Would you
   - Write the report as directed by the boss?
   - Refuse to write the report, because to do so would be unethical?
   - Write the report, but also write a memo to the boss stating that what was being done was unethical–to cover you in case you are found out?
   - Write the report as directed, but refuse to have your name on it as the author?
   - Go over your boss's head and report that you have been asked to falsify records?
   - Do something else? (Please specify)

Use the *Evaluation Checklist, Ethics Check List,* and the **Five P's** to help you answer these questions. (From *Chemical Engineering Magazine,* p.132, September 1980)

8. In the case above, Jay has written the report to suit his boss, and the company has gone ahead with an ambitious commercialization program for Catalyst A. Jay has been put in charge of the pilot plant where development work is being done on the project. To allay his doubts, he personally runs some clandestine tests on the two catalysts. To his astonishment and dismay, the tests determine that while Catalyst A works better under most conditions (as everyone expected), at the operating conditions specified in the firm's process design, Catalyst B is indeed considerably superior.
    If you were Jay, what would you do?
    *   Since no one knows that you've done the tests, do you just keep quiet about them since the process will run acceptably with Catalyst A, but not nearly as well?
    *   Do you tell your former boss (the catalyst expert) about the clandestine tests and let him decide what to do next?
    *   Do you make clean breast of the whole affair to higher management, knowing that it could get you and a number of colleagues fired or, at least, discredited professionally?
    *   Do something else? (Please specify.)
    Use the *Evaluation Checklist, Ethics Check List,* and the **Five P's** to help you answer these questions. (From *Chemical Engineering Magazine,* p.132, September 1980)

9. Ruth works as a group leader for a company that sells large quantities of a major food product that is processed before sale, by heating. Her product-development group has been analyzing the naturally occurring flavor constituents of the product and she discovers that several of the flavor components (actually pyrolysis products, present in minute quantities) are chemicals that have been found to cause cancer in animals when given in large doses. Yet, the product— in worldwide use for literally centuries— has never been implicated as a cause of cancer.
    Although in the U.S. the Delaney Amendment to the Pure Food and Drug Act prohibits adding cancer-causing agents to food, there are no government regulations concerning those that may occur naturally.
    Should Ruth
    *   Quash the report?
    *   Submit a confidential report to her superiors and let them decide?
    *   Submit an article summarizing the compounds found to a reputable journal, but without mentioning cancer?
    *   Notify a consumer-protection organization.
    Use the *Evaluation Checklist, Ethics Check List,* and the **Five P's** to help you answer these questions. (From *Chemical Engineering Magazine,* p. 132, September 1980)

10. The company that employs Reginald has a practice of using salaried personnel to replace the striking workers and to pay these people double-time pay for any work over 40 hours per week, plus a $100-per-day strike bonus. (Under ordinary circumstances, overtime pay is never granted to salaried personnel, which includes engineers.)
    Not having a union themselves, Reginald and his fellow engineers have been hard hit by inflation, and many welcome the opportunity to earn extra pay.
    The plant is presently being struck by union operators over "unsafe" working conditions, which Reginald personally believes *may* be unsafe but which are not covered specifically under government safety regulations. The company disputes the union's contention about safety. The strike looks as if it could be a lengthy one.

Should Reginald
- Refuse to work, because he thinks the union's allegations may have merit?
- Refuse to work, because he believes that strikebreaking is unethical?
- Work, because he feels this is an obligation of all members of management?
- Work, because the extra pay is a great way to catch up on some of his bills, or earn the down payment on a car, etc.?
- Work, because he believes he may be fired if he doesn't?
- Do something else? (Please specify)

Use the *Evaluation Checklist, Ethics Check List,* and the **Five P's** to help you answer these questions. (From *Chemical Engineering Magazine*, p. 132, September 1980)

11. Larry's company has been using a flavor additive in one of its products, but there have been problems with the flavor's stability. One of Larry's chemists accidentally finds the flavor can be stabilized by adding a mixture of tin and lead in very small quantities. Although both tin and lead are recognized poisons, the chemist points out the amounts added are not more than might be leached out of the soldered seams of the common tin cans used for a multitude of food products. The new product will be packed in glass, so no further addition of heavy metals will occur.
    Should Larry
    - Recommend that the additive not be used, because it is unethical to add poisons no matter what the quantity?
    - Prevent any further problems by suppressing the finding?
    - Recommend the open use of this heavy-metals-stabilized additive?
    - Recommend that it be used, but that the deliberate addition of heavy metals be considered a trade secret, and be kept from leaking to the public because "it would only cause unnecessary worry?"

    Use the *Evaluation Checklist, Ethics Check List,* and the **Five P's** to help you answer these questions. (From *Chemical Engineering Magazine,* p. 132, September 1980)

12. Ken is a process engineer for Stardust Chemical Corp., and he has signed a secrecy agreement with the firm that prohibits his divulging information that the company considers proprietary. Stardust has developed an adaptation of a standard piece of equipment that is highly efficient for cooling a viscous plastics slurry. (Stardust decides not to patent the idea but to keep it a trade secret.) Eventually, Ken leaves Stardust and goes to work for a candy-processing company that is not in any way in competition. He soon realizes that modification similar to Stardust's trade secret could be applied to a different machine used for cooling fudge, and at once has the change made.
    Consider
    - Has Ken acted unethically (because he divulged the proprietary modification that Stardust developed)?
    - Has Ken acted ethically (because he has only used the idea behind the modification, and not the specific change developed by Stardust)?
    - Would Ken have acted unethically if the machine used to cool fudge and the one used by Stardust were identical?

    Use the *Evaluation Checklist, Ethics Check List,* and the **Five P's** to help you answer these questions. (From *Chemical Engineering Magazine,* p. 132, September 1980)

13. You put together a buoyancy exhibit for a "science day" event. The exhibit consists of filling up an aquarium with water, then testing which soda pop cans float. The idea is that the diet soda cans float because they don't contain sugar, while the nondiet soda pop cans sink. In preparing your experiment the night before, you discover that one of your diet soda pop cans floats but the other slowly sinks to the bottom. You know that if you "spike" the water with salt, increasing the water density, all the diet soda pop cans will float. Do you

- Spike the water and tell no one?
- Spike the water and tell only those parents who appear to be looking at this as a future science fair experiment?
- Spike the water, but put up a sign next to the exhibit warning people about what you have done?
- Not spike the water?
- Do something else? (Please specify)

Use the *Evaluation Checklist, Ethics Check List,* and the **Five P's** to help you answer these questions. (Contributed by Dr. Susan Montgomery, University of Michigan, 1992)

14. You attend your friend's piano recital, knowing that your friend is seriously considering a career as a pianist. Her performance is atrocious. She approaches you after the performance and asks, "Well, what did you think?" Do you
    - Tell her to keep her day job?
    - Tell her it sounded terrific?
    - Try to get out of it by telling her Victor Borge couldn't have done a better job?
    - Tell her it was pretty good, with plans to approach her later to discourage her from making a career of it?
    - Do something else? (Please specify)

    Use the *Evaluation Checklist, Ethics Check List,* and the **Five P's** to help you answer these questions. (Contributed by Dr. Susan Montgomery, University of Michigan, 1992)

15. Your company sends you to a foreign country to take part in the bidding for a large construction project. Once you get there, your associates in the country tell you the bribing rates for the officials in charge of taking the bids. Do you
    - Refuse to bribe the officials, knowing you will certainly lose the project?
    - Bribe them, double the ongoing rate, to ensure that you get the project?
    - Go along with the bribing, knowing that it is the only way your project will be considered?
    - Do something else? (Please specify)

    Use the *Evaluation Checklist, Ethics Check List,* and the **Five P's** to help you answer these questions. (Contributed by Dr. Susan Montgomery, University of Michigan, 1992)

16. Your boss takes you out on the town dining the first weekend of your summer job, not knowing that you are still under the legal drinking age. In the restaurant, everyone at your table orders a drink.
    Do you
    - Order a beer, wanting to fit in, and hope they don't check ID?
    - Order a beer, wanting to fit in, ready with your false ID in case they ask for one?
    - Order a fancy nonalcoholic drink, so you can still sort of fit in, even though you hate them?
    - Order your favorite soft drink?
    - Do something else? (Please specify)

    Use the *Evaluation Checklist, Ethics Check List,* and the **Five P's** to help you answer these questions. (Contributed by Dr. Susan Montgomery, University of Michigan, 1992)

8 *PUTTING IT ALL TOGETHER*

In the previous seven chapters we have presented the building blocks of creative problem solving individually in order to focus more easily on each block. When faced with a problem, we bring many principles of the different building blocks to bear on the problem at the same time. We have shown the heuristic in the form of a job card below. Job cards are usually displayed at the workplace or carried around by individuals to help recall previously learned concepts that can be applied to their job.

JOB CARD

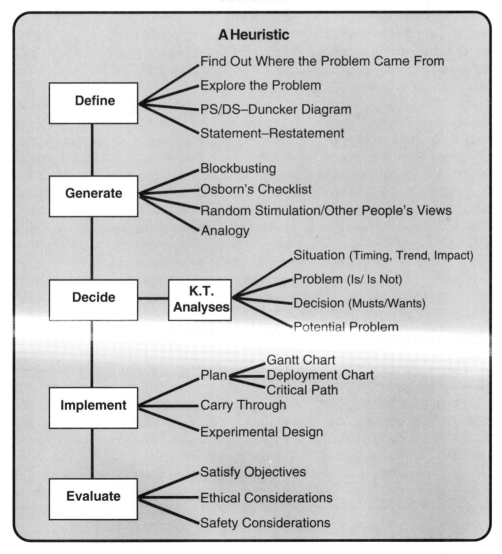

A Heuristic

Define
- Find Out Where the Problem Came From
- Explore the Problem
- PS/DS–Duncker Diagram
- Statement–Restatement

Generate
- Blockbusting
- Osborn's Checklist
- Random Stimulation/Other People's Views
- Analogy

Decide — **K.T. Analyses**
- Situation (Timing, Trend, Impact)
- Problem (Is/ Is Not)
- Decision (Musts/Wants)
- Potential Problem

Implement
- Plan
  - Gantt Chart
  - Deployment Chart
  - Critical Path
- Carry Through
- Experimental Design

Evaluate
- Satisfy Objectives
- Ethical Considerations
- Safety Considerations

We will now show how the principles we have been studying can be applied to two case studies.

8.1 CASE STUDY: MEET ME AT THE MALL (OR NO MORE BIG BOOK MAIL ORDERS)[†]

Sears, Roebuck and Company marked the end of an era in 1992 when they announced elimination of their catalog sales. Additionally, Sears announced that they would close more than 100 unprofitable stores in the face of a fourth quarter 1992 deficit of $830 million (their first quarterly loss in nearly 60 years). What led to this situation?

The catalog dates back to 1886 when Robert W. Sears, a railroad station agent, began selling watches and jewelry with mailers. Nine years later, with Sears Watch Company employee Alvah Roebuck, they produced their first general merchandise catalog containing 532 pages of merchandise aimed at America's farm families. The catalog quickly caught on and became a lifeline for people in remote areas, as well as a piece of American folklore.

People living in rural America kept abreast of the latest developments through the catalog. Over the years, the catalog became an illustrated encyclopedia, a reflection of the technology and fashions of the times. By 1992, the Sears catalog offerings had grown to include Spring/Summer, Fall/Winter, the Great American Wish Book Christmas catalogs, and 50 other seasonal, monthly, and specialized catalogs, each of which was mailed to over 14 million households. The catalogs posted sales of $3.3 billion in U.S. sales in 1992. A great success story? Yes, and no. The catalog division was still in serious trouble, experiencing significant red ink in spite of the seemingly impressive sales. Why?

Sears management clearly failed to recognize and respond to a paradigm shift. They had the attitude that "We can sell anything we want as long as it says Sears on it." They failed to respond to a fundamental shift in the nature of the American public over the years, from a primarily agrarian economy to the high-tech 1990s. While other direct marketers were producing slick, targeted niche catalogs to capitalize on customer's specific interests, Sears was still producing a bloated, costly "Big Book" that featured unprofitable items such as camping equipment and major appliances. These items had not been successful catalog sellers for years. (Would anyone in the 1990s consider purchasing stoves and refrigerators or washers and dryers from a catalog?) America had changed, and Sears had failed to change with it. This was a case of continuing to do things in a certain way because that's the way they had always been done. Classic paradigm paralysis.

Two and one half years earlier, J.C. Penney's catalog division took in 197 million phone calls, yet Sears would not acknowledge that phone orders were important and didn't even install an (800) number until 1992. As one industry executive put it, Sears was "stuck in concrete" and far back on the learning curve.

[†]Direct Marketing, p.6, March 1993.

The ripple effect from the elimination of the Big Book hurt more than Sears and Sears's customers:

R.R. Donnelley Printers	–lost 60% of business in Chicago plant (closed) lost 50% of business in Elgin, Ill. plant (810 workers affected)
Paper Companies	–lost 40,000 tons/yr in business
U.S. Postal Service	–lost $100 million/yr in postage and handling

Let's analyze this scenario and apply some problem-solving techniques to determine alternative courses of action that Sears might have taken (with 20/20 hindsight).

First, why was the problem allowed to get so serious before action was taken? We don't know for a fact how long this problem was developing, but from the information regarding the type of products still shown in the catalog and the lack of efficient phone sales, this looks like an example of paradigm paralysis. Sears had blinders on that prevented them from realizing that the direct marketing industry was changing around them. The Sears catalog continued "business as usual" until it was too late. Their competition had overtaken and surpassed them. Suppose that we were faced with the same situation while there was still time to make corrections. What kind of ideas can we come up with?

Problem Definition

Let's use a Duncker Diagram to clarify the problem definition.

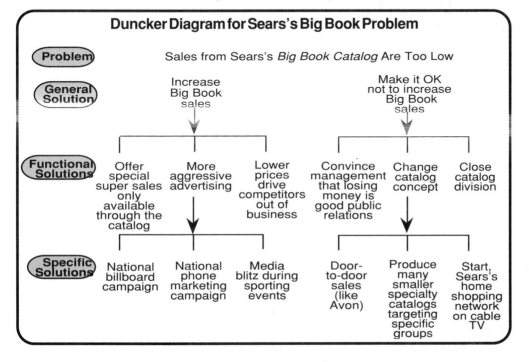

Duncker Diagram for Sears's Big Book Problem

| Problem | Sales from Sears's *Big Book Catalog* Are Too Low |

General Solution: Increase Big Book sales / Make it OK not to increase Big Book sales

Functional Solutions: Offer special super sales only available through the catalog / More aggressive advertising / Lower prices drive competitors out of business / Convince management that losing money is good public relations / Change catalog concept / Close catalog division

Specific Solutions: National billboard campaign / National phone marketing campaign / Media blitz during sporting events / Door-to-door sales (like Avon) / Produce many smaller specialty catalogs targeting specific groups / Start, Sears's home shopping network on cable TV

Another problem definition technique that we could try is the *Statement-Restatment* technique. Let's apply trigger 3:

<u>Trigger No. 3:</u> Make an opposite statement:

Opposite Statement to Having Good Catalog Sales	New Avenues Suggested by Opposite Statement
• Raise the price so high nobody will buy anything.	• Are our prices competitive?
• Don't let anybody see the catalog.	• Are we circulating the catalogs sufficiently, but without being wasteful?
• Print the catalog in a foreign language so nobody can read it.	• Are the catalogs easy to read? Maybe we ought to consider a big print version for the elderly.
• Treat customers poorly so that they are angered and don't buy anything.	• Is the service adequate in the catalog division? Have we done customer satisfaction surveys?
• Don't answer the phone, or leave it off the hook so it's always busy when customers call.	• Are the phone lines adequate? (an 800 number?) Maybe we should provide computer modem access to an on-line catalog.
• Sell only one high-priced unpopular item.	• Diversify the catalog offerings.

We can separate the thoughts generated by the trigger into two specific categories:

1) Assessing customer satisfaction with the catalog division
   - competitive prices?
   - adequate service?
   - adequate phone service?

While this category contains concerns important for any retail business, and should be addressed, these thoughts probably do not get to the root of the problem with the catalog division. We must consider what's unique about catalog sales and how we can target improvement in those areas. Consider the second category:

2) Improvement of the catalog marketing strategies
   - Is the catalog attractive and easy to read?
   - Is the catalog distributed widely to potential customers?
   - Can the catalog offerings be diversified?

Some of these ideas tie back to the points on the Duncker Diagram. As we reflect on these ideas we can come to the conclusion that we ought to try to improve sales in the catalog division through more efforts than just the Big Book. Our redefined problem statement can thus be stated:

> **How can we improve sales in the catalog division through more efforts than just the Big Book?**

It is clear from this statement that the problem should be attacked on a variety of "fronts." Some possible "fronts" were developed through the trigger and the Duncker Diagram. These include a national billboard campaign, national phone marketing, a home shopping network, the production and specialty catalogs targeting specific groups, toll free phone lines, diversification of offerings, and competitive prices.

Generate Solutions

Using the Duncker Diagram and the Statement-Retatement technique (triggers), we have already generated some ideas on how to solve the problem. Let's summarize them at this point with a fishbone diagram.

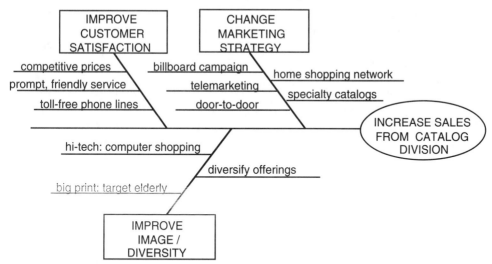

Now let's brainstorm some possible alternatives for one of the solutions we've suggested. Consider the specific solution:

> **POSSIBLE SOLUTION**
>
> **Produce many smaller specialty catalogs targeting specific groups.**

Some ideas that occur immediately are a tool catalog, lawn and garden catalog, and toy catalog.

We can continue with this procedure and probably generate a list of departments in a typical store. Let's try to use some of our techniques to stimulate new ideas.

Other People's Views

Babies/Small Children
- baby clothes marketed to expectant mothers
- baby shower catalogs
- educational games for preschool
- specialty baby foods from around the world
- handy gadgets for baby care

Elderly
- convenience items for the infirm
- large print books
- travel agency in a catalog... cruises/bus tours

Teenagers
- team logo products
- sound equipment
- high-tech computer games (virtual reality)
- CD's /Tapes
- MTV videos

Newlyweds
- honeymoon packages, cruises–travel agency
- housing locators/national real estate service
- gift ideas

This technique generated some ideas that we hadn't thought of previously. One interesting "niche" idea that might be worth pursuing is that of a national catalog travel agency that could target trips and tours for such diverse groups as newlyweds and the elderly. In addition, the entertainment market as a whole seemed to generate a number of possibilities as well: CD's, tapes, video/computer games, educational games, sound equipment, large print books, etc.. These hot areas may be fertile ground for diversification and individually targeted catalog offerings. Let's try another of the brainstorming techniques, random stimulation, and see what we come up with.

Random Stimulation–to generate additional specialty catalog ideas

Choose a word from our random word list in Chapter 4.

The randomly chosen word *violin* brings to mind
music → CD's/Tapes/Music Videos
instrument → musical instruments → marching band accessories for schools
strings → fishing → fishing equipment → boating accessories, nautical items, nautical fashions, other specialty fashion clothes
bow → archery → uncommon sporting goods items

Try another randomly chosen word: *Ghost* brings to mind

<u>witches</u> → brooms → cleaning products/appliances/gadgets
<u>white</u> → sheets → specialty bedding items/towels, etc.
<u>Halloween</u> → costumes

Again, more entertainment and leisure ideas were generated. Additionally, several new ideas surfaced: fashions (specialty fashions and costumes), convenience items, and high-tech gadgets.

We can summarize the ideas that we have generated thus far with another fishbone diagram.

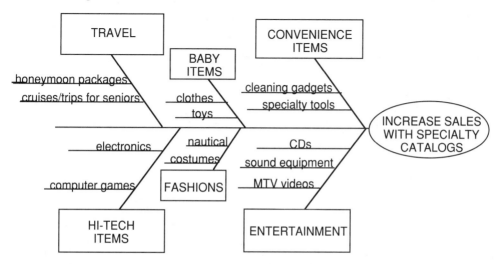

We have narrowed the problem solution down to producing a series of targeted individual niche catalogs in one or more of the following areas:

- Travel
- High Tech Gadgets
- Specialty Fashion
- Baby Accessories
- Entertainment
- Convenience Items

Each catalog would list the (800) telephone number, should be customer accessible by computer, and be staffed by trained, courteous service personnel.

Our targeted marketing approach should be concerned with how to gain access to the appropriate customer base for each of our new catalog lines. When we decide upon which catalogs to implement, we must be sure that we have an efficient targeted advertising/distribution plan. Perhaps our marketing efforts could be supplemented with television advertising and a toll free (800) phone number that customers could use to request individual catalogs. The television advertising itself could be targeted to specific times (children's items on Saturday morning cartoons, sporting goods/entertainment items during sporting events).

The next step in our solution process would be to move on to a K.T. Decision Analysis. We first construct a decision statement.

| **Decision Statement:** | Determine which of the individual catalogs we should print and distribute. |

We then list our musts and wants for our new catalog lines and then make some decisions. (i.e., We *must* show a profit in less than 18 months. We *must* be able to selectively target the appropriate customers for each individual catalog. (If we send each catalog to everyone, it would be more costly than the Big Book distribution.) We *want* to be able to operate the new catalog without hiring a lot of new personnel. And so on... .) In conjunction with this solution process, we would also have to perform a Potential Problem Analysis to warn us of any possible complications inherent in our solution.

The decision was made to launch a travel agency advertising catalog targeted at arranging trips to foreign countries. We now construct a Potential Problem Analysis table for this individual service catalog.

Potential Problem Analysis Table for Travel Agency Catalog

Potential Problem	Possible Cause	Preventive Action	Contingent Action
No one signs up for tour offerings	Poor advertising, wrong season	Targeted mailings, sharp brochures	TV commercials, discounted fares
Customer gets sick in foreign country	Food poisoning, virus or bacterial infection	Education on eating habits to observe during trip	Have names of English speaking doctors in area, fly passenger home on commercial airline
Obnoxious traveler makes trip miserable for co-travelers	There's one in every crowd	Pass out brochure on travel etiquette to each customer, train tour guides to deal with difficult people	Refund the problem traveler's money and fly him home on commercial airline, punch him out

Once this Potential Problem Analysis has been performed, and we have assured ourselves that we have planned for potential problems, we would then evaluate the solution and implement our chosen course of action.

Careful analysis of the problem and the alternative solutions may have shown that the best thing for Sears to have done was to close the catalog division. If indeed that is the case, something else that could be examined would be how to make the best of a bad situation and still perhaps profit from the closing? For example, could they have sold their customer list to other direct marketers and recoup some funds in that manner? This example illustrates the uses of a variety of our techniques on an actual problem, and a chance to practice them one more time.

Another example that exercises Problem Analysis and Decision Analysis skills is the **Coors Brewery** case study.

8.2 CASE STUDY: THE SILVER BULLET™ OR... CARRY ME BACK TO OLD VIRGINY*

The Situation: A major brewery in the Rocky Mountains, Coors, was running at maximum capacity and could not meet the supply demands in the East. In addition, the shipping costs of the finished product to the East reduced the profits in spite of a somewhat higher selling price in the East. Consequently, the company considered the possibility of building a new plant where it would be well positioned to reach both the northeastern and southeastern markets.

Whether or not to build a new plant. First, it had to be decided whether or not Coors should take a risk and invest millions of dollars to build a second plant to increase capacity. The company was quite profitable and clearly this venture was not without associated risks, especially if the water supply in the new plant could not produce the same taste as that produced in the Rocky Mountains. If the venture failed, the company's good name would be put at risk. Being a family run company with its only facility in the Rocky Mountains, it could be very difficult to control quality and operations at a far away site. Additionally, a marketing problem would be sure to surface even if the new plant's waters were acceptable, because this clearly could negate 35 years of marketing strategy claims that the water used in the original plant in the Rocky Mountains is the primary reason for the beer's good taste. However, the infrastructure and environmental constraints were such that the Rocky Mountain plant could not produce any more beer. Eventually the company *did* decide to take a risk and invest several million dollars to build a second plant.

Where to build? The next question was where to build the plant. There are two sets of criteria: one set that *must* be satisfied and another set that they would *like* to satisfy. The choices narrowed to building a second plant in one of four locations. While the actual analysis involved a number of other criteria in both the *musts* and

K.T.
Decision
Analysis

---

*Scenario developed from an article in *Enterprise*, Vol. 7, No.3, p. 45, 1994 and discussions with Terry Zinsli, Coors.

the *wants*, the following reconstruction of a K.T. Decision Table does provide a number of the issues involved in the decision.

The *musts* were acceptable water quality, easy accessiblity to both rail and highway, adequate number of people in the area to staff the plant, and availability of land with high quality water.

Musts	Virginia	Site 2	Site 3	Site 4
Water Quality	GO	GO	GO	NO GO
Sufficient Workforce	GO	GO	GO	GO
Transportation Access to Plant	GO	GO	GO	GO
Land Available	GO	GO	GO	NO GO

Wants	Wt	Rating	Score	Rating	Score	Rating	Score	Rating	Score
Local Business Environment	6	8	48	5	30	4	24	NO GO	
Quality of Water	10	8	80	5	50	5	50		
Quality of Workforce	8	7	56	8	64	6	48		
Vicinity of Major Eastern Market	8	8	64	7	56	3	24		
			248		200		146		

Because the Virginia site received the highest score, we will make it our tentative choice and look at the adverse consequences.

Virginia Adverse Consequences			
Consequence	Probability of Occurance	Seriousness	PXS
1) Perception of alcohol in the region.	4	4	16
2) Poor marketing strategy because beer does not contain Rocky Mountain spring water.	10	10	100

Problem Definition

After choosing Virginia, serious concern remained about the Rocky Mountain water marketing strategy. The company could either change the marketing strategy and possibly risk losing the major share of the current market or perhaps they could do something about the water problem.

Let's find out *where the problem came from*. Was the reasoning valid that the Rocky Mountain plant was really at maximum capacity and not able to produce any more beer? We need to *gather information* about the process. Were all the steps running at their maximum capacity or were there one or two steps limiting the output? (i.e., what was the bottleneck?) *Explore the problem* by identifying and examining the input and output rates to each processing step and determining which steps are at maximum capacity. Calculations on a simplified version of the problem showed that there are sufficient raw materials (inputs: water, malt, hops, and other ingredients) to more than double the capacity. The first steps in the processing are mixing steps and the production capacity could certainly be increased in these steps. However, after gathering more information, it was determined that the capacity of the "finishing steps" (e.g., packaging, waste stream discharge) could just not be increased from a practical standpoint in the existing plant. Consequently, a new problem statement was formulated as: "How can we increase the capacity of the *finishing steps* and meet the supply demands of the eastern market?"

| Generate Solutions |

A number of possible solutions are
1) change environmental contraints.
2) increase the number of units in the final stages of the current plant.
3) ship the unfinished beer to Virginia for final processing and packaging.

More information was collected and an engineering and economic analysis was carried out on a number of the possible solutions. First, there was no space available to add more units for the finishing process. Secondly, there was concern about handling increased waste from the finishing steps in the Rocky Mountain site. However an analysis showed that it would be economically and technically feasible to ship the unfinished beer by rail in tank cars to Virginia for final processing. With this solution, the company could still increase capacity and lay claim to the fact that it was the Rocky Mountain water supply that gave the beer its unique flavor and taste. Thus, by challenging the original assumptions, that all the steps were slow and that the finished product had to be made at one site, innovative solutions could be obtained.

| Gather Information |

A plant was built in the Shenandoah Mountains of Virginia and opened in 1987. The plant capacity was designed for 50,000 barrels per week and within a couple of years the demand was exceeding this amount and a new problem arose: How to increase the capacity of the Virginia plant.

Increasing the Capacity of the Virginia Plant. In analyzing the situation, it was found that the output at the Virginia plant varied at certain times of the year. In mid-summer the plant was able to increase its output to meet higher demands while in the winter months it could not. A K.T. Table could be constructed in the following manner to analyze the new problem of variable capacity during the year.

| Problem Definition |

	IS	IS NOT	DISTINCTION
What	Limited output of finished beer.	Limited supply of unfinished beer.	One or more steps in the finishing process may limit output.
Where	East	West	Processing steps affected differently in East than West.
When	Winter	Summer	Higher Temperatures in East during summer.
Extent	Temperature affects unit exposed to seasonal temperature variations.	Units not affected by season.	Waste pond exposed to seasonal in temperature variations.

Generate Solutions

The higher temperature in the summer was linked to increased bacterial activity in the sludge pond which allowed the waste streams from the plant to be processed more rapidly. A new problem statement would be: *Find a way to increase the waste pond capacity during the winter months.* Possible solutions are 1) add a second waste pond, 2) heat the waste pond, and 3) heat the air that was bubbled through the pond during processing. The best solution was to heat the air bubbling up through the waste pond. When this solution was implemented, along with computer control of the inventory, the capacity of the Virginia plant doubled with no additional investment of capital.

Implement

CLOSURE–GO FORTH AND SOLVE PROBLEMS

Solving problems can be fun and add spice to an otherwise bland existence. The techniques discussed in this book for defining and solving real-world problems are tried and proven and can make you a better problem solver. Use the materials in this book to lessen your problem-solving anxiety.

- Take a proactive approach not only to problem solving but also to your life.

- Have a vision of what you want to accomplish in your organization and in your life and make sure all the steps you take are in the right direction.

- Don't be afraid to fail. No one bats a thousand and most of the time we can learn more from our failures than we do from our successes.

- Observe good problem solvers. Ask them questions and brainstorm with them.

- Don't dismiss ideas of others out of hand, leverage these ideas–they may trigger the correct solution.

- Get in the habit of planning your time and prioritizing the tasks you must accomplish.

- Practice using the tools and techniques presented in the book. A tool only becomes useful when you are comfortable and familiar with it. Some techniques may sound silly or cumbersome. Don't worry. The same technique will not work for *everyone* and *every* situation. Choose what works best for you.

- Don't force-fit a certain technique that you have used before. If it's not working, choose another or move on.

- When you have finished a task, reflect on it. Evaluate it. Is it the best you could have done? If not, can you still improve upon it?

Every time you practice the problem-solving techniques we have presented, you become an even better problem solver.

APPENDIX 1

McMaster Five-Point Strategy[*]

1. *Define*:
   a. Identify the unknown or stated objective.
   b. Isolate the system and identify the knowns and unknowns (inputs, laws, assumptions, criteria, and constraints) stated in the problem.
   c. List the inferred constraints and the inferred criteria.
   d. Identify the stated criteria.

2. *Explore*:
   a. Identify tentative pertinent relationships among inputs, outputs, and unknowns.
   b. Recall past related problems or experiences, pertinent theories, and fundamentals.
   c. Hypothesize, visualize, idealize, generalize.
   d. Discover what the real problem and the real constraints are.
   e. Consider both short-time and long-time implications.
   f. Identify meaningful criteria.
   g. Choose a basis or a reference set of conditions.
   h. Collect missing information, resources, or data.
   i. Guess the answer or result.
   j. Simplify the problem to obtain an "order-of-magnitude" result.
   k. If you cannot solve the proposed problem, first solve some related problems or solve part of the problem.

Developing A Strategy

3. *Plan*:
   a. Identify the problem type and select among heuristic tactics.
   b. Generate alternative ways to achieve the objective.
   c. Map out the solution procedure (algorithm) to be used.
   d. Assemble resources needed.

4. *Act*:
   a. Follow the procedure developed under the plan phase; use the resources available.
   b. Evaluate and compare alternatives.
   c. Eliminate alternatives that do not meet all the objectives or fulfill all the constraints.
   d. Select the best alternative of those remaining.

5. *Reflect*:
   a. Check that the solution is blunder-free.
   b. Check reasonableness of results.
   c. Check procedure and logic of your arguments.
   d. Communicate results.

[*]Woods, D.R., *A Strategy for Problem Solving*, 3rd ed., Department of Chemical Engineering, McMaster University, Hamilton, Ontario, 1985; *Chem. Eng. Educ.*, p. 132, Summer 1979; *AIChE Symposium Series*, 79 (228), 1983.

APPENDIX 2

PLOTTING DATA

In this appendix, we review the techniques for plotting data and measuring slopes on various types of graphs. With the aid of many readily available computer packages, constructing graphs from data is quite straightforward. The background material in this appendix will help you understand the various types of graphs and enable you to determine the important parameters from them. Additionally, we review some statistical techniques available for analyzing experimental data.

A2.1 Linear Plots

First, let's look at a quick review of the fundamentals of graph construction and slope measurement on linear plots. Equations of the form

$$y = mx + b \qquad\qquad\qquad\text{(A-1)}$$

will, of course, yield a straight line when plotted on linear axes. Consider an example where we place $100 in the bank in a simple interest bearing account for five years. A graph of the amount of money in the account at the end of each year is shown below.

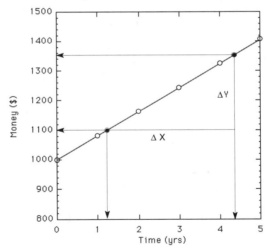

To determine the slope of the line on the graph, we could proceed in two ways.

Method 1–Direct Measurement

We can physically measure Δx and Δy using a ruler, and then using the linear scale of the graph, determine the slope of the line. For this example, suppose $\Delta x = 34$ mm and $\Delta y = 20$ mm. The linear scale for the y-axis is $12.50/mm and for the x-axis it is (1yr)/(11 mm). Therefore,

$$\Delta Money = (20 \text{ mm}) (\$12.50/\text{mm}) = \$250$$

$$\Delta Time = (34 \text{ mm}) (1\text{yr})/(11 \text{ mm}) = 3.1\text{yrs}$$

$$Slope = \frac{\Delta Money}{\Delta Time} = \frac{\$250}{3.1 yrs} = \$81 / yr$$

The intercept ($1000) can be read directly from the graph at time t = 0. While this method works, it is rather crude and its accuracy is limited by the accuracy of the measuring devices used.

Method 2–Direct Calculation from Points on the Line

Pick two points that lie on the line. These points will not necessarily be data points if the data are scattered.

Point 1:   (1.23 , 1100)
Point 2:   (4.32 , 1350)

$$Slope = \frac{\Delta Money}{\Delta Time} = \frac{1350 - 1100}{4.32 - 1.23} = \$80.9 / yr$$

The intercept may be determined from either point using the calculated slope and the equation of the line ($y = mx + b$, $1100 = (80.9)(1.23) + b$, $b = 1000$). As you can see there is a slight difference in the values of the slopes calculated using the two different methods. This difference can be attributed to the accuracy with which the point values can be read from the graphs and the accuracy of the measurements in Method 1.

Many computer packages are available which can determine the equation of the "best" straight line through the data points. These programs employ a statistical technique called regression (or least-squares) analysis. A regression analysis of the above data yields the following equation for the "best" line through the data points:

$$Money(\$) = [\$81/ yr] [Time(yrs)] + 1000$$
$$y \quad = \quad m \qquad x \qquad + \quad b$$

Notice the values determined in this manner are very close to those determined earlier.

A2.2 Log-Log Plots

When both axes are logarithmic, the graph is called a log-log graph. A log-log graph of the data is used when the dependent variable (u, for instance) is proportional to the independent variable (say, v) raised to some power m:

$$u = Bv^m \qquad (A\text{-}2)$$

In many engineering applications, it is necessary to determine the best values of m and B for a set of experimental measurements on u and v. One of the easiest ways to perform this task is to use logarithms on Equation A-2. If we take the log of both sides of the equation, we get

$$\log u = m \log v + \log B \qquad (A\text{-}3)$$

Now, if we let

$$y = \log u$$
$$x = \log v$$
$$b = \log B$$

then Equation (A-3) becomes:

$$y = mx + b \qquad (A\text{-}4)$$

Now we can now clearly see that Equation (A-2) has been transformed so that a plot of $\log u$ (y) versus $\log v$ (x) will be a straight line with a slope of m and an intercept of $\log B$. Chemical reaction rate data often follow a log-log relationship. Consider the following reaction rate data:

Clearly nonlinear!!

Concentration, C_A (gmoles/dm^3)	1	2	3	4
Reaction Rate (gmoles/dm^3/hr)	3	12	27	48

If we assume that a log-log plot (Equation (A-2)) is appropriate here, we can graph these data to determine m and B. There are two ways that we can proceed. We can manually take logarithms of the data and plot those, or we can use log-log coordinates and let the graph do the work. We shall illustrate both methods.

Method 1: Manually taking logarithms (Note: $\log = \log_{10}$ in this example)

log(Concentration)	0	0.301	0.477	0.602
log(Reaction Rate)	0.478	1.08	1.43	1.68

We now plot these points on linear paper. It is very important to remember that if you manually take logs, you must plot the points on **linear** paper *not* log-log paper (otherwise you'll get a mess!).

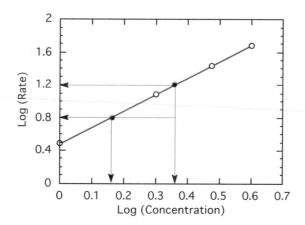

Now we can proceed just as we did for the simple linear plot case to determine the slope and intercept of the line and find the parameters for Equation (A-2). The intercept, b, is clearly 0.478 (from the tabular data). Let's determine the slope using two points on the line.

Point 1:　(0.38, 1.2)
Point 2:　(0.18, 0.8)

$$Slope = \frac{1.2 - 0.8}{0.38 - 0.18} = m$$

and since $b = 0.478$,

$$b = \log B = 0.478$$

$$B = 10^b = 10^{0.478} = 3$$

Thus, the equation for the line is

$$u = Bv^m = 3v^2$$

or in terms of concentration and rate

$$Rate = 3C_A^2$$

Method 2: Plotting Directly on Log-Log Paper

　　　Plotting directly on log-log paper is relatively simple. You merely plot the points, and the logarithmic scales on the axes take the logs for you. The rate versus concentration data are plotted below on log-log axes.

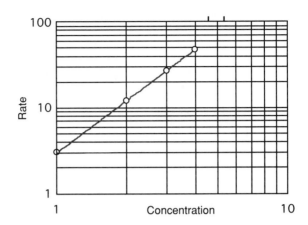

Notice that we again obtain a straight line. These types of plots are more meaningful to the reader, because the points that are plotted correspond to the actual numerical values of the data, and not the logarithms of the data. From the plot, the reader can determine immediately that the concentration value associated with the second data point is 2 gmole/dm^3. This physical intuition is not available when the logarithms of the data are plotted on a linear scale (e.g., we would know the log of the concentration is 0.303, but we wouldn't know the concentration itself, unless we could perform antilogs in our heads). So now how do we determine the slope and "intercept" from this type of a plot? If Equation (A-3) is a valid equation for the line, then it should hold for every point on the line. Writing this equation for two arbitrary points on the line, we get

$$\log u_1 = \log B + m \log v_1 \qquad \text{Point 1} \qquad (A\text{-}5)$$

$$\log u_2 = \log B + m \log v_2 \qquad \text{Point 2}$$

If we subtract the equation for point 2 from that of point 1, we'll get an expression that will allow us to calculate the slope of the line on the log-log axes.

$$\log u_1 - \log u_2 = m(\log v_1 - \log v_2)$$

$$m = \frac{\log(\frac{u_1}{u_2})}{\log(\frac{v_1}{v_2})} \quad \text{or, for this example } m = \frac{\log(\frac{\text{rate 1}}{\text{rate 2}})}{\log(\frac{C_{A1}}{C_{A2}})}$$

Once the value of m is determined, B can be found by substituting the appropriate values for either point back into Equation (A-5).

For this example,

$$m = \dfrac{\log(\dfrac{48}{3})}{\log(\dfrac{4}{1})} = \dfrac{1.204}{0.602} = 2$$

$$\log B = \log 48 - 2\log 4 = 0.477$$

$$B = 3$$

Notice that this is the same result that we arrived at previously, as it should be.

We can also determine the slope of a line on log-log axes using the direct measurement technique, as discussed in the section on linear plots. With the availability of computer graphing packages this method is not used very much, but we include it here for completeness. To measure the slope directly using a ruler or similar instrument, we would choose two points on the line, and measure Δx and Δy and the cycle length in both directions and proceed as follows.

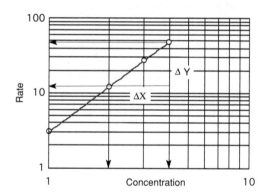

$\Delta y = 2.55$ cm
Cycle length in the y–direction = 3.6 cm/cycle

$\Delta x = 1.55$ cm
Cycle length in the x–direction = 4.35 cm/cycle

$$\left. \begin{array}{l} \Delta y = \dfrac{2.55 \text{ cm}}{3.6 \text{ cm / cycle}} \\[2em] \Delta x = \dfrac{1.55 \text{ cm}}{4.35 \text{ cm / cycle}} \end{array} \right] \Rightarrow \dfrac{\Delta x}{\Delta y} = \dfrac{2.55 * 4.35}{1.55 * 3.60} = 2.0$$

therefore, $$y = bx^2$$

and, using the point, $(x = 1, y = 3)$, we can determine the value of b:

$$3 = b(1)^2, \quad b = 3$$

$$y = 3x^2$$

$$\text{rate} = 3C_A^2$$

Again, this is the same result we obtained earlier.

A2.3 Semi-Log Plots

Semi-logarithmic (semi-log) plots should be used with exponential growth or decay equations of the form

$$y = be^{mx} \quad \text{or} \quad y = b(10)^{mx} \tag{A-6}$$

To determine the parameters, b and m, we take logarithms of both sides of Equation (A-6). We'll use the "e" form of the equation and natural logarithms (ln) , although the result is the same for the other form of the equation using common (base 10) logarithms.

$$\ln(y) = \ln b + mx \tag{A-7}$$

Examining Equation (A-7) we see that a plot of $\ln y$ versus x should be a straight line with a slope of m and an intercept of b. If we deposit $100 into a bank account that gathers interest compounded continuously (a great deal!), and then plot the amount of money in the account at the end of every year for the first ten years, we obtain the following graph:

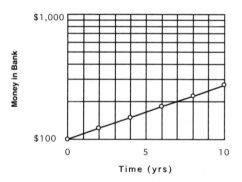

Notice that we have plotted the data (for every other year) directly on semi-log paper. We could have manually taken the logs and plotted the data on linear paper as discussed earlier, but using semi-log axes is easier. We show two methods to determine the parameters of Equation (A-6) from the semi-log plot.

Method 1: Algebraic Method

Draw the best straight line through your data points. Choose two points on this line and determine the x and y values of each point.

Point 1: (8 yrs, $223)   Point 2: (2 yrs, $122)

Notice that we can write Equation (A-7), using the two points and solve for the slope.

$$\ln(y_1) = \ln b + mx_1$$

$$\ln(y_2) = \ln b + mx_2$$

$$\ln(\frac{y_2}{y_1}) = m(x_2 - x_1)$$

$$m = \frac{\ln(\frac{y_2}{y_1})}{(x_2 - x_1)}$$

Substituting the values of the selected points, we get $m = \dfrac{\ln\left(\dfrac{223}{122}\right)}{8-2} = 0.1$, and

then b is clearly the y value of the line at time t = 0, i.e., $be^{0.1(0)} = b$, thus $b = 100$

$$\text{Amount}(\$) = \$100e^{0.1\text{Time(yrs)}}$$

Method 2: Graphical Technique

A modification of the algebraic method is possible on semi-log paper if we extend the "best" line we can draw so that the dependent variable, y, changes by a factor of 10. For this case, the ratio of y_2/y_1 is 10 and the equation for the slope of the line is merely:

$$m = \frac{\ln(10)}{x_2 - x_1} = \frac{2.303}{x_2 - x_1}$$

The intercept can then be determined as before. This technique is referred to as the *decade method.* Careful analysis and plotting of the data are important tools in problem solving. In addition to being able to calculate the slopes and intercepts, we should be able to deal with "scatter" in the data. The next section discusses this topic.

A 2.4 Establishing Confidence Limits for Data

Normally, the data we wish to analyze are the results of some type of experiment. There will often be some "scatter" or variability in the data. If an experiment is repeated a number of times, chances are the results will have some variation. These variations are due to experimental error, instrument precision, material variability, etc. One approach to dealing with this situation is to perform several experimental runs under identical conditions and average the results. Intuitively, we realize that the more runs we perform, the closer the average of our experiments will approach the true average for the experimental conditions under consideration. In fact, if we performed an infinite number of repetitions, we would expect to obtain the true average as a result of this hard work. Since we cannot afford to perform an infinite number of experiments to determine the true mean of these experiments (μ = true mean) we would like to have a way to estimate it from a limited number of samples. Let's define the following quantities:

μ = true mean of the experiments (i.e., population) if a very large (∞)
    number are performed

n = sample size (the number of experiments that we actually performed)

$$\bar{x} = \text{mean of the } n \text{ samples} = \frac{\sum x}{n}$$

The standard deviation of the samples quantifies the spread of the sample values about the sample mean. The bigger the standard deviation, the larger the variability of the individual samples. We can estimate the standard deviation of the entire population (from which we have measured n samples) using Equation (A-8):

$$S(x) = \sqrt{\frac{\sum x^2 - n\bar{x}^2}{n-1}} \qquad (A-8)$$

Note, to determine the true standard deviation (σ) of the entire population, we would have to know μ, the true mean (which we don't). So, we estimate the standard deviation.

As an example of a distribution, consider the following. We could gather some data regarding the length of time beyond the bachelor's degree that it takes a student at Frostbite Falls University to complete a Ph.D.. If we plotted the data, it might look like the following figure.

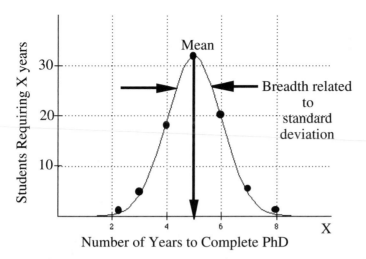

Number of Years to Complete PhD

The mean time to obtain a Ph.D. is five years, and there is clearly a distribution about the mean, which is to be expected. Almost no students take less than two years or longer than eight years. The breadth of this "distribution" is related to the standard deviation.

If we draw several different samples of size n from the entire population, and calculate the mean for each, we will most likely get a different mean, $\overline{x}$, for each sample. Additionally, none of these means will probably be equal to the true mean of the entire population, μ. In other words, we will get a distribution of means that is related to the true mean μ. The variability of these means is related to the variability of the entire population (i.e., to the standard deviation).

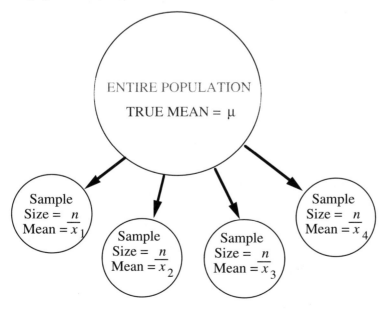

Let $S(\bar{x})$ represent the estimated standard deviation of the means of samples of size n drawn from the entire population which has an estimated standard deviation of $S(x)$ (that we previously calculated).

These quantities are related in the following manner:

$$S(\bar{x}) = \frac{S(x)}{\sqrt{n}} \qquad \text{(A-9)}$$

We noted that $\bar{x}$ is an estimate of the true mean (μ) of the experiments. But how good an estimate is it? One way to express this relationship is

$$\mu = \bar{x} \pm t\,S(\bar{x}) \qquad \text{(A-10)}$$

This equation states that the true mean lies within $t\,S(\bar{x})$ of the estimated mean, $\bar{x}$. The quantity t is calculated from the so-called t distribution. t is a function of two parameters: a confidence level and the degrees of freedom. The degrees of freedom in our context is $(n-1)$ where n is the number of experiments performed under a given set of conditions. For a set of experiments, the number of degrees of freedom that we have to model the data is n, the number of runs. For this case, we have used one degree of freedom to calculate the mean. Thus there are only $(n-1)$ degrees of freedom remaining. Another way to look at this is that we can specify $(n-1)$ data points independently, and then the nth value is fixed, since we have already determined the mean value $\bar{x}$ of the data set. The confidence level is defined as the probability that t is smaller than the tabulated value. Below is an excerpt from a table of the t distribution:

Degrees of Freedom	95% Confidence Level	99% Confidence Level
1	12.706	63.657
2	4.303	9.925
3	3.182	5.841
4	2.776	4.604
5	2.571	4.032
6	2.447	3.707
7	2.365	3.499
8	2.306	3.355
9	2.262	3.250
10	2.228	3.169

In essence, the confidence level will tell us how certain we can be that the mean of our data points will lie within the calculated range. For example, with a 95% confidence level, the means will lie within the range in 95% of the cases, and outside the range in 5% of the cases.

An example will help here. Consider an experiment where we pop popcorn using a certain method. We make five runs under identical conditions and obtain the following data:

Run No.	% Unpopped Kernels of Popcorn
1	22
2	18
3	19
4	26
5	25

(Population Size)
$$n = 5$$

(Estimated Mean)
$$\bar{x} = \frac{\Sigma x}{5} = \frac{110}{5} = 22$$

(Estimated standard deviation of the individual points from the estimated mean)

$$S(x) = \sqrt{\frac{\Sigma x^2 - n\,x^2}{n-1}} = \sqrt{\frac{2470 - (5)(22)^2}{5-1}} = 3.5355$$

(Estimated standard deviation of the sample means from the true mean)

$$S(\bar{x}) = \frac{S(x)}{\sqrt{n}} = \frac{3.5355}{\sqrt{5}} = 1.581$$

Now to estimate the true mean, we'll use Equation (A-10):

$$\mu = \bar{x} \pm t\,S(\bar{x}) = 22 \pm 1.581t$$

For four degrees of freedom ($n - 1 = 4$), we can be 95% confident that $t \le 2.776$ (see the table of t values). Similarly, we can be 99% confident that $t \le 4.604$. Hence,

$$\mu = 22 \pm 1.581\,(2.776) = 22 \pm 4.39 \quad \text{(with 95\% confidence)}$$

or

$$\mu = 22 \pm 1.581\,(4.604) = 22 \pm 7.28 \quad \text{(with 99\% confidence)}$$

Note that the more confidence with which we want to specify the mean, the larger the "error bars" become, or the larger n must be for the same size uncertainty. Data points can be plotted using these values to indicate the uncertainty in the measurement.

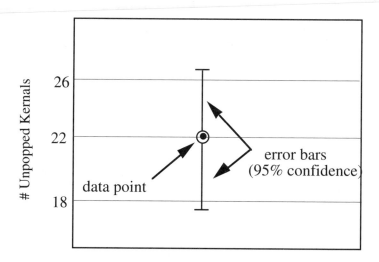

More extensive listings of the *t* distribution can be found in numerous texts on statistics.

References: Volk, William, *Applied Statistics for Engineers*, 2nd ed., McGraw-Hill, New York, 1969.
Glantz, Stanton A., *Primer of Biostatistics*, 3rd ed., McGraw-Hill Health Professions Division, New York, 1992.

A2.5 Presentation of Data

The data that we gather, when properly organized, analyzed and presented will help serve as the basis for subsequent decision making. In order to be of maximum use for problem-solving purposes, organization and presentation of the information are very important. Drawings, sketches, graphs of data, etc. can all be effective communication tools when used properly. Try to display and analyze the data in such a manner so as to extract meaningful information. When presented with data, analyze it to make sure it has not been biased to lead you in the wrong direction.

Display the data graphically rather than in tabular form. Tables can be difficult to interpret and sometimes terribly misleading, as demonstrated by Anscombe's Quartet[1,2] shown on the next page. Graphing, on the other hand, is an excellent way to organize and analyze large amounts of data.

From a statistical standpoint (using a regression analysis on the tabular information) all the data sets are described equally well by the same linear model. However, upon graphing the data, we see some very obvious differences. The graphical presentation (on the next page) clearly reveals the differences in the data sets, which may have gone unnoticed if we had used only the tabular data.

[1]Tufte, E., *Visual Display of Quantitative Information*, Graphics Press, Cheshire, CT, 1983.
[2]Anscombe, F. J., "Graphics in Statistical Analysis," *Amer. Statisticians*, 27, p. 14-21, Feb. 1973.

Set A		Set B		Set C		Set D	
X	*Y*	*X*	*Y*	*X*	*Y*	*X*	*Y*
10.0	8.04	10.0	9.14	10.0	7.46	8.0	6.58
8.0	6.95	8.0	8.14	8.0	6.77	8.0	5.76
13.0	7.58	13.0	8.74	13.0	12.74	8.0	7.71
9.0	8.81	9.0	8.77	9.0	7.11	8.0	8.84
11.0	8.33	11.0	9.26	11.0	7.81	8.0	8.47
14.0	9.96	14.0	8.10	14.0	8.84	8.0	7.04
6.0	7.24	6.0	6.13	6.0	6.08	8.0	5.25
4.0	4.26	4.0	3.10	4.0	5.39	19.0	12.50
12.0	10.84	12.0	9.113	12.0	8.15	8.0	5.56
7.0	4.82	7.0	7.26	7.0	6.42	8.0	7.91
5.0	5.68	5.0	4.74	5.0	5.73	8.0	6.89

Anscombe's Quartet Table

Each of these four data sets A, B, C, and D all have the following properties:

$N = 11$ Mean of X's = 9.0   Equation of regression line: $Y = 3 + 0.5\,X$

$t = 4.24$ Mean of Y's = 7.5   Standard error of estimate of slope = 0.118

$r^2 = 0.67$ Correlation coefficient = 0.82   Sum of squares $(X - \overline{X})^2 = 110.0$

Regression sum of squares = 27.50   Residual sum of squares of $Y = 13.75$

Statistically, Everything *Looks* the Same !!

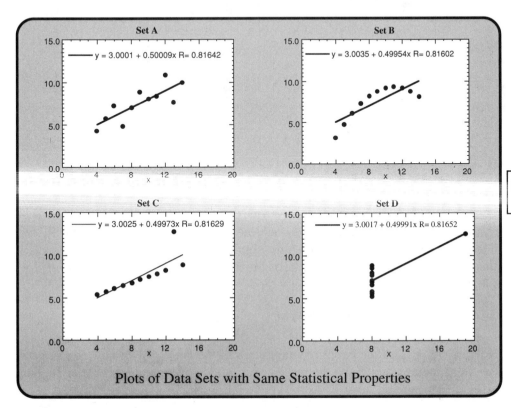

Everything *Looks* Different !!

Plots of Data Sets with Same Statistical Properties

A2.6 Getting the Most Out of Your Data: Grouping the Variables

Sometimes, a particular quantity that we are interested in measuring is dependent upon a number of different variables. Let's consider the flow of fluid through a pipe, at low velocities. Under special low-flow conditions the flow is termed laminar. We perform a number of experiments and determine that the pressure drop through the pipe depends upon a number of different parameters: the diameter of the pipe, the density of the liquid, the velocity of the liquid and the viscosity of the liquid. (The viscosity is a physical property of the liquid. It is related to how "thick" the liquid is. For example, maple syrup is more viscous than water, and molasses is more viscous than maple syrup.) We can graph the data that we have obtained on separate graphs. That is, we can graph how pressure drop varies with fluid velocity, while we hold all the other variables constant. Then, we could graph pressure drop versus pipe diameter, while holding everything else constant. In this way we would generate a series of four individual graphs.

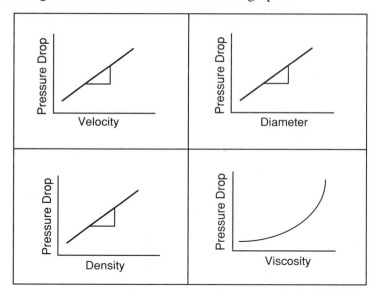

However, if we carefully analyze the physics of the situation, we may be able to determine a grouping of the variables that would provide us with all the information that we have available in a single graph. The variable that serves this function for fluid flow through a pipe is the Reynolds Number, which is defined as follows:

$$\text{Reynolds Number} = \frac{(\text{Density})(\text{Velocity})(\text{Diameter})}{(\text{Viscosity})}$$

Now, if we process the experimental data into the form of pressure drop versus Reynolds number and graph the results, we find that we obtain a single graph that gives us the same information as the above four graphs.

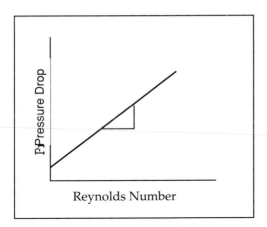

Proper selection of the grouping of the variables will not necessarily be known ahead of time, but if you are aware of this possibility, you may be able to condense your data into a more usable format.

INDEX